Fundamentals and Applications of Renewable Energy

Fundamentals and Applications of Renewable Energy

Stephen Adams

STATES
ACADEMIC PRESS
www.statesacademicpress.com

Fundamentals and Applications of Renewable Energy
Stephen Adams
ISBN: 978-1-63989-221-1 (Hardback)

STATES
ACADEMIC PRESS

Published by States Academic Press,
109 South 5th Street,
Brooklyn, NY 11249, USA

Cataloging-in-Publication Data

Fundamentals and applications of renewable energy / Stephen Adams.
p. cm.
Includes bibliographical references and index.
ISBN 978-1-63989-221-1
1. Renewable energy sources. 2. Power resources. I. Adams, Stephen.
TJ808 .F86 2022
333.794--dc23

For more information regarding States Academic Press and its products, please visit the publisher's website www.statesacademicpress.com

Table of Contents

Preface

The energy collected from renewable resources such as sunlight, tides, waves and wind is called renewable energy. These energy sources are carbon neutral in nature and are replenished naturally on a human timescale. Some areas of major importance under this discipline are heating and cooling systems, electricity generation, and transportation. A few commonly used technologies for utilizing renewable energy sources are hydropower systems, wind turbines, bioenergy systems, solar panels and geothermal energy solutions. Renewable energy systems are expected to reduce air pollution and lower the dependency on conventional sources of fuel. This book is compiled in such a manner, that it will provide in-depth knowledge about the theory and practice of renewable energy. It picks up individual sources of energy and explains their need and contribution in the context of a growing economy. As this field is emerging at a rapid pace, the contents of this book will help the readers understand the modern concepts and applications of the subject.

Given below is the chapter wise description of the book:

Chapter 1- Renewable energy is defined as the useful energy that is produced or collected from renewable sources that are naturally and constantly replenished. These natural sources include solar energy, wind energy, hydro energy, geothermal energy, biomass energy and tidal energy. This chapter introduces the reader to renewable energy and its socio-economic impact. **Chapter 2**, Wind energy is the process of harnessing the power of the flowing wind or airflows that occur naturally in the Earth's atmosphere to create electricity. The simplest and most efficient method is a wind turbine. A wind turbine is a mechanically powered wind device that turns electrical generators for electrical power output. This chapter covers all aspects related to wind energy and its utilization. **Chapter 3**, Solar energy is the energy that is generated from the Sun in the form of radiant light and heat. The process of conversion of this energy from the Sun is known as solar power. Solar thermal energy is the collection of the Sun's heat for human use. This chapter discusses all the topics related to solar energy like solar thermal power plants, active solar, solar power, etc. in a comprehensive manner. **Chapter 4**, Hydropower is the power generated from the energy of fast-running water that can be used for various purposes. When this power is converted to electricity, through the use of generators, it is called hydroelectricity. This chapter has been carefully written to provide the reader with a better understanding of hydropower. **Chapter 5**, Biomass energy refers to the generation of energy by organic materials. Biomass is any plant material used as a source of fuel. Biofuel and biogas are two renewable energy sources derived from organic wastes like food scraps, animal wastes, feedstock, etc. Biomass energy and its features are explained in detail in this chapter. **Chapter 6**, Geothermal energy is the thermal energy stored and generated inside the earth. Thermal energy refers to the energy that determines the temperature of the matter in the Earth's crust. Geothermal energy is of two types – vapor-dominated form and liquid-dominated form. This chapter focuses on geothermal energy and its various applications.

Indeed, my job was extremely crucial and challenging as I had to ensure that every chapter is informative and structured in a student-friendly manner. I am thankful for the support provided by my family and colleagues during the completion of this book.

Stephen Adams

1

Introduction to Renewable Energy

Renewable energy is defined as the useful energy that is produced or collected from renewable sources that are naturally and constantly replenished. These natural sources include solar energy, wind energy, hydro energy, geothermal energy, biomass energy and tidal energy. This chapter introduces the reader to renewable energy and its socio economic impact.

Renewable power is booming, as innovation brings down costs and starts to deliver on the promise of a clean energy future. American solar and wind generation are breaking records and being integrated into the national electricity grid without compromising reliability.

Wind turbines and a large solar panel.

This means that renewables are increasingly displacing "dirty" fossil fuels in the power sector, offering the benefit of lower emissions of carbon and other types of pollution. But not all sources of energy marketed as "renewable" are beneficial to the environment. Biomass and large hydroelectric dams create difficult tradeoffs when considering the impact on wildlife, climate change, and other issues.

Renewable energy, often referred to as clean energy, comes from natural sources or processes that are constantly replenished. For example, sunlight or wind keep shining and blowing, even if their availability depends on time and weather.

While renewable energy is often thought of as a new technology, harnessing nature's power has long been used for heating, transportation, lighting, and more. Wind has powered boats to sail the seas and windmills to grind grain. The sun has provided warmth during the day and helped kindle fires to last into the evening. But over the past 500 years or so, humans increasingly turned to cheaper, dirtier energy sources such as coal and fracked gas.

Now that we have increasingly innovative and less-expensive ways to capture and retain wind and solar energy, renewables are becoming a more important power source, accounting for more than one-eighth of U.S. generation. The expansion in renewables is also happening at scales large and small, from rooftop solar panels on homes that can sell power back to the grid to giant offshore wind farms. Even some entire rural communities rely on renewable energy for heating and lighting.

As renewable use continues to grow, a key goal will be to modernize electricity grid, making it smarter, more secure, and better integrated across regions.

Dirty Energy

Nonrenewable, or "dirty," energy includes fossil fuels such as oil, gas, and coal. Nonrenewable sources of energy are only available in limited amounts and take a long time to replenish. When we pump gas at the station, we're using a finite resource refined from crude oil that's been around since prehistoric times.

Nonrenewable energy sources are also typically found in specific parts of the world, making them more plentiful in some nations than others. By contrast, every country has access to sunshine and wind. Prioritizing nonrenewable energy can also improve national security by reducing a country's reliance on exports from fossil fuel–rich nations.

Many nonrenewable energy sources can endanger the environment or human health. For example, oil drilling might require strip-mining Canada's boreal forest, the technology associated with fracking can cause earthquakes and water pollution, and coal power plants foul the air. To top it off, all these activities contribute to global warming.

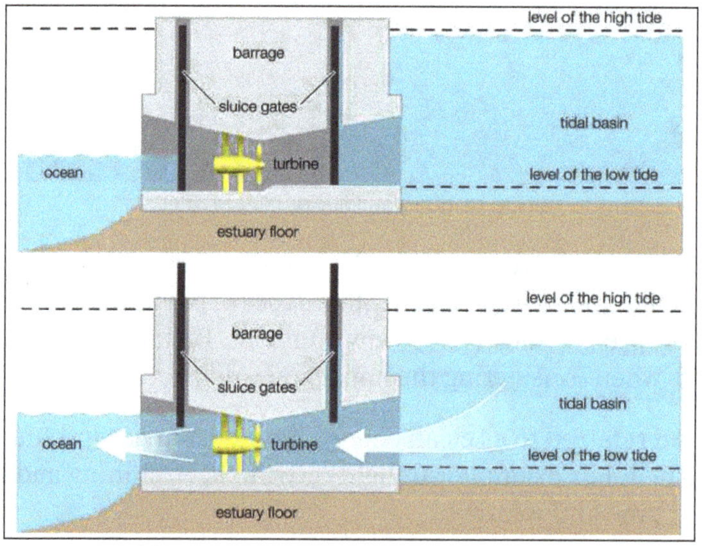

Tidal power: Diagram of a tidal power barrage.

At the beginning of the 21st century, about 80 percent of the world's energy supply was derived from fossil fuels such as coal, petroleum, and natural gas. Fossil fuels are finite resources; most estimates suggest that the proven reserves of oil are large enough to meet global demand at least until the middle of the 21st century. Fossil fuel combustion has a number of negative environmental consequences. Fossil-fueled power plants emit air pollutants such as sulfur dioxide, particulate matter, nitrogen oxides, and toxic chemicals (heavy metals: mercury, chromium, and arsenic), and mobile sources, such as fossil-fueled vehicles, emit nitrogen oxides, carbon monoxide, and particulate matter. Exposure to these pollutants can cause heart disease, asthma, and other human health problems. In addition, emissions from fossil fuel combustion are responsible for acid rain, which has led to the acidification of many lakes and consequent damage to aquatic life, leaf damage in many forests, and the production of smog in or near many urban areas. Furthermore, the burning of fossil fuels releases carbon dioxide (CO_2), one of the main greenhouse gases that cause global warming.

In contrast, renewable energy sources accounted for nearly 20 percent of global energy consumption at the beginning of the 21st century, largely from traditional uses of biomass such as wood for heating and cooking. By 2015 about 16 percent of the world's total electricity came from large hydroelectric power plants, whereas other types of renewable energy (such as solar, wind, and geothermal) accounted for 6 percent of total electricity generation. Some energy analysts consider nuclear power to be a form of renewable energy because of its low carbon emissions; nuclear power generated 10.6 percent of the world's electricity in 2015.

Cost Effectiveness

The ultimate question for the policy strategist is whether power generated from a renewable energy source is affordable given the service it is providing.

Does Renewable Energy Generate Affordable Power?

On a total cost basis, a new, renewable energy, generating facility is often cost competitive with a conventional fuel facility provided that the cost calculation considers long-term fuel costs - and even more so when one considers environmental costs and benefits. Since this generalization is not true in every situation confronting the policy planner, the policy planner will need to apply cost-effectiveness criteria adapted to each situation.

What are the Applicable Cost-effectiveness Evaluation Criteria?

Any given electric generating technology (including renewables) may be cost effective in one market or application and not in another. There is no simple calculus a policy maker can apply, but a number of established criteria will assist in determining the financial viability of renewable energy generation.

The Quality and Quantity of the Resource

Quality and quantity of renewable resources may be determined by a government-conducted resource assessment, but private-sector developers commonly have their own pre-feasibility and feasibility studies which can be more accurate measures of the commercial viability of a given project. The measures of resource quality and quantity are unique to each resource, but for each of the renewables, resource quality and quantity affects the energy input to, and the effective capacity of a generation facility. In geothermal resource development, for example, the temperature of the resource and the dissolved impurities determine the requisite production equipment. The cost of production equipment, in turn, affects the installed cost and the per-kilowatt-hour cost of delivered power. In biomass, the quality and BTU content of the fuel will influence installed costs, and operations and maintenance costs.

The Location of the Resource

Proximity of a resource to a customer base directly impacts costs, as does proximity to an existing infrastructure (roads, transmission lines, etc.), to industry support facilities (concrete plants, etc.), and to the developer's technology manufacturing base. In the case of geothermal, the depth of the resource is a major cost factor. For the hydro, wind and solar technologies, climatic variations (rainfall, cloud cover, intense storms) affect cost. For biomass, transportation distances between the fuel source and the generating facility may significantly affect the electricity cost.

Government-imposed Costs

For the private-sector developer, time is money. The time expended in responding to bid proposals, in obtaining requisite permits, licenses and concessions, and in negotiating contracts increases the costs of renewable projects. The policy maker should consider policies that organize and simplify the local institutional processes. Such policies may prevent adding major costs and time delays to what would otherwise be a highly cost-effective facility. Similarly, government-imposed taxes, fees, tariffs and royalty payments are all passed to the electricity consumer and affect the kilowatt-hour cost of delivered power.

The development process is sufficiently similar that for many purposes the renewable resources may be treated similarly. The development process employed by hydroelectric, geothermal, wind energy, biomass and solar renewable technologies may be described and analyzed in three discrete stages: *"reconnaissance"*, *"exploration"* and *"exploitation"*. Each technology may use different terms for these three stages, but the concepts are similar.

- Reconnaissance is an activity which determines by visual observation and scientific studies whether an area may be a source of commercially exploitable resources. It does not affect the present surface use of the land.
- Exploration is an activity which demonstrates the dimensions, position, characteristics and extent of resources by scientific studies. It may affect the present surface use of the land. An extensive outlay of capital may be required for exploring the potential of some renewable resources such as geothermal and, to a lesser degree, wind and hydro resources. Where resource exploration is expensive, this may necessitate that exclusive rights to the relevant renewable resources in the area be awarded to the explorer. These rights may be granted for a limited term, but if exploration proves the commercial viability of the resource, a private-sector developer will require that the temporary exploration rights be converted to long-term exploitation rights.
- Exploitation is an activity which enables electricity to be produced from renewable resources, either through the intermediate production of steam or the direct production of electricity from a chemical or mechanical process. Exclusive, long-term rights are prerequisite to the sustainability of this production.

The expense associated with each phase of resource development has a direct impact on the cost-effectiveness of the electricity produced. The first generation facilities developed in any resource area are almost always going to cost more per kilowatt hour produced than will later facilities, since most of the reconnaissance and exploration costs will be included in the cost of the first facility. If, however, there is some assurance of a market for power from additional facilities, should the initial facility prove feasible, the reconnaissance and exploration costs may be allocated over more kilowatt hours thus reducing the initial cost? By allocating initial reconaissance and exploration costs over multiple projects, the per kilowatt hour cost may be significantly reduced.

Financing Costs

For renewables, the bulk of a project's total lifetime cost is represented by the initial capital cost, and will be incurred before the project ever comes on line. The cost of a renewable energy is in the technology effort exerted at the outset of a project and all of the renewables share *"front-end-loaded"* cost profiles. Consequently, the majority of new generation facilities are funded through project financing, whereby the principal and interest (and profit) are paid from the proceeds of the project.

Cost of Electricity

Power Purchase Agreement

The "power purchase agreement" - the power and capacity contract between the owner of a generating facility and its customers - rather than the credit-worthiness of the developer, collateralizes the loan. Since renewable energy projects are front-end-loaded, the cost of capital significantly affects installed cost. High risk factors associated with initial projects developed in new resource areas also translate into higher costs of capital. The challenge for the country policy maker - especially in a country which seeks to attract initial projects to a new resource area - is to implement new mechanisms to lower financing costs. Development of such mechanisms may prove more productive if done in consultation with the private-sector developer. For example, in some situations, municipal customers may have access to tax-exempt or low interest bonds that can be used to finance energy projects at a lower cost than if they were financed with conventional borrowing.

In a rapidly advancing technological era, the most prudent course for a decision maker is to avoid reliance on old information as to whether a given renewable technology can fulfill a given energy need.

System Costs

The cost or cost-savings of integrating a given renewable energy generator into a system is difficult to quantify. By diversifying the energy supply mix, a system can protect or buffer the ratepayer from the potential financial risks and volatility of changing fuel prices, changing environmental requirements, and common design flaws that can result in large operational and maintenance costs. Reliance on imported fuels may be eliminated and balances of payment problems thereby reduced. With the exception of biomass, there are no intrinsic fuel costs for an established renewable energy generating facility. Consequently, an established renewable facility serves as a hedge against inflation in an inflationary market.

For an example of a renewable facility as an inflationary hedge one may examine the history of the older hydroelectric power dams. The following chart illustrates the renewable technologies which are currently available in the marketplace.

Important Characteristics			
	Options	Status	Capacity
Small hydro	Low to high head turbines and dams. Run of river.	Virtually all are commercial.	Factor Intermittent to base load.
Wind	Horizontal and vertical axis wind turbines. Wind Pumps.	Commercial. New designs under development.	Variable, 20 to 40%.
Solar	Photovoltaic. Active thermal (low to high temp for heat or electricity). Passive thermal.	Most commercial. Some under development or refinement.	W/o storage: <25%, intermittent. W/thermal storage: 0 to 60%, intermediate.
Geothermal	Cycles: Dry steam, Flash, and Binary	Commercial. Exploration and drilling improvements underway.	High, base load.
Bioenergy	Combustion. Fermentation. Digestion. Gasification. Liquefication.	Many commercial. More under development or refinement.	US wood plants average 95+%. Intermediate, peaking also possible.

Important Characteristics of the Renewable Technologies

- All six renewable energy sectors offer technologies which are proven and are available in the marketplace. All can be purchased today in forms that are reliable and cost-competitive.
- "Capacity Factor" summarizes the output patterns:
    - Geothermal and most biomass plants provide baseload energy.
    - Most hydro and some biomass plants are highly dispatchable, offering a range of options from baseload to peaking.
    - Run-of-river hydro is intermittent, but variations in its output tend to be slow and predictable.
    - Solar ranges from intermittent to intermediate, depending on how well it matches the pattern of energy usage.
    - Wind is intermittent, but studies have found that most grids can add an intermittent source up to 15% of their capacity without requiring any compensatory action. Higher shares from intermittent sources are usually easy to accommodate.

Socio-economic Impact

Social Impacts

These resources also provide social benefits like improvement of health, according to choice of consumer, advancement in technologies, and opportunities for the work, but some basic considerations should be taken for the benefit of humans, for example, climate conditions, level of education and standard of living, and region whether urban or rural from agricultural point of view. Social aspects are the basic considerations for the development of any country. The following social benefits can be achieved by renewable energy systems: local employment, better health, job opportunities, and consumer choice. The study concluded that the total emission reduction is exponentially increasing in different years after the installation of renewable energy projects in remote areas. Social impacts of each resource with its magnitude are listed in table.

Table: Social impacts assessment for different renewable energy sources.

Technology	Impact	Magnitude
Photovoltaic	Toxins	Minor-Major
	Visual	Minor
Wind	Bird strike	Minor
	Noise	Minor
	Visual	Minor
Hydro	Displacement	Minor-Major
	Agricultural	Minor-Major
	River damage	Minor-Major
Geothermal	Seismic activity	Minor
	Odor	Minor
	Pollution	Minor-Major
	Noise	Minor

Economics

It was discovered that renewable energy projects provide benefits in economic point of view because they utilize local labor from rural areas, local material and business, local shareholders, and services of local banks. In addition, the renewable energy projects have facilitated the communities by establishing a trust fund that aims to invest the money earned by selling electricity in local economy. This makes it easy for a few communities to invest money on any small business of their own choice. Biofuel projects created large number of jobs; however, very low jobs were created by solar power plants, as the ratio of people working in different company's increase that will create more jobs for others by using the part of their economy for entertainment, leisure, restaurant, etc. The consumers will be provided with electric power at a low cost as compared to that of conventional energy sources, and overall economy will be enhanced because there will be multiple options to generate power using different renewable energy sources present in that region.

Environmental Impacts

Renewable energy projects have also contributed in improving environmental impacts such as reduction of carbon dioxide gas, awakening community about the climate change. The study observed very small impacts on the people living in a particular area, tourism, cost of energy supply, and educational impacts. Significant impacts were observed in improvement of life standard, social bonds creation, and community development. They also observed that the renewable energy projects are complex to install and are local environmental and condition sensitive. Their forecasting, execution, and planning require more consideration and knowledge as compared to other projects. The two main aspects of environment are air and water pollution, normally created by the discharged water from houses, industries, and polluted rain, and discharge of used oils and liquids contains poisonous chemicals and heavy metals like mercury, lead, etc. Along with water pollution, natural resources can be maintained and greenhouse effect and air pollution can be mitigated by the proper usage of renewable energy sources as shown in table. Carbon dioxide emission with the generation of electric power using different energy resources is given in figure.

Table: Summary of environmental effects.

Category of impact	Relationship to conventional sources	Comment
	Exposure to harmful chemicals.	
Emission of Hg, Cd, and other toxic elements.	Reduced emissions.	Emission reduced a few hundred times.
Emission of particles.	Reduced emissions.	Much less emission.
	Exposure to harmful gases.	
CO_2 emission.	Reduced emissions.	A big advantage.
Acid rain, SO, NO_x.	Reduced emissions.	Reduced more than 25 times.
Other greenhouse gases.	Reduced greenhouse gases.	Big advantage-global warming.
	Other	
Spouts off fossil fuels.	Total or partial elimination of oil spills.	Heavy fuel oil and other petroleum product spills.
Water quality.	Better quality water.	Reduced water pollution.
Soil erosion.	Smaller loss of land.	In most cases, there is no penetration deep into earth.

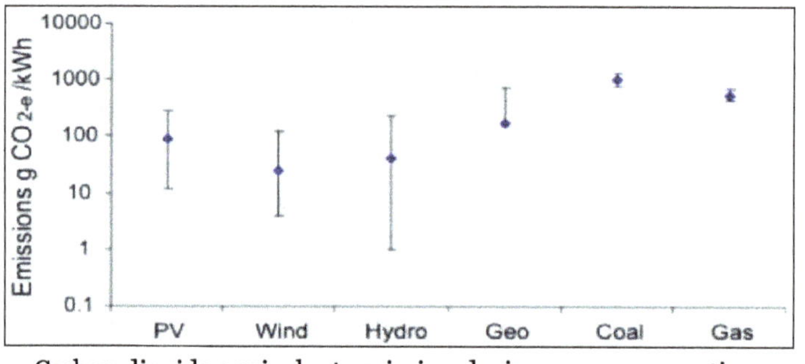

Carbon dioxide equivalent emission during power generation.

Various greenhouse gases in atmosphere are being increased by humankind by doing many economic activities. The role of greenhouse gases and current situation are given in table.

Table: Role of different substances in greenhouse effect.

Substance	Ability to retain infrared radiation compared to CO_2	Preindustrial concentration	Present concentration	Annual growth rate (%)	Share in greenhouse effect due to human activity	Share in greenhouse effect increase due to human activity
Alpha	1	275	346	0.4	71	50 ± 5
Beta	25	0.75	1.65	1.0	8	15 ± 5
Gamma	250	0.25	0.35	0.2	18	9 ± 5
Delta	17,500	0	0.00023	5.0	1	13 ± 5
Epsilon	20,000	0	0.00040	5.0	2	13 ± 5

Sociopolitical Impacts

Solar panels are usually installed at the roofs of the buildings that increase the job opportunities in the PV system fabrication and installation. This increases the regional development and reduces the usage of energy from nonrenewable energy projects. It is very useful at the regions where there is no access of electricity. The major problem with solar system is the high investment and maintenance cost. Biomass energy projects have great contribution in the local job creation and the development of rural areas. Such types of power plants have large opportunities of jobs in construction of plants, management, and maintenance of plants, production, and preparation of biomass. Only the noise production and unpleasant smell are the negative impacts of these plants. Fuel cells have slow implementation because of their high cost of plant construction and energy generation. Their construction and operation create jobs in almost all technical activities. In hydro power plants, the major sociopolitical problem is the shifting of the people from the areas where the plant is going to be constructed. These plants provide significant jobs for local community and also play an important role in the economic development of the community. The construction of tidal energy plants has no effect on humans, and they have better contribution in the local and official employment. These plants are very expensive and are not common. Wind energy projects do not have any emigration problem, and they create large number of job opportunities especially for engineers. Geo thermal energy projects provide the following sociopolitical benefits: improvement in the education of local people, improvement in living standards, and improvement in the care of health issues.

Impacts on Grid

When the solar panels are connected to the distribution system, the cost of safety equipment is reduced because their short circuit current is higher than the nominal value. Biomass power plants have the same effects on the gird as do conventional plants. The integration of wind energy plants, tidal energy, and geothermal energy is complex.

Socioeconomic Impacts

Three case studies were made to investigate the socioeconomic benefits of renewable energy projects, and the three cases were solar, wind, and biofuel energy projects; empirical method was used to collect data. The basic aim of study was to know the contribution of renewable energy projects to local sustainability, which includes social, economic, and environmental, and to identify the socioeconomic benefits of REPs through the concerned community. It was done by doing survey of the communities. Eleven parameters were used including job creation, impacts on education, easy usage of energy, income development, demographic impacts, social bonds creation and community development, usage of native resources, and tourism. They concluded that the impacts of REPs on employment are positive, and indirect employment is high in comparison with the size of community, whereas direct employment is moderate.

Availability and Technical Limitations

One of the important assessing factors to generate power from renewable energy sources is the availability and their technical limitation. Each resource has some limitations; photovoltaic has limitation to generate power only because heat energy from sun can only be received during the day time, except cloudy season. For wind turbine, speed should not increase beyond 25 m/s; otherwise, turbine will be damaged. Also, low speed of wind, that is, <3 m/s, will not be sufficient for the generation of electric power. Geothermal has good ability to generate power throughout the day for 24 h but is geography limited according to the presence of resources. Hydro-electric power plants are easy to start, stop, and operate within minutes; hence, they are considered as one of the highest available, reliable, and flexible renewable energy resources. From efficiency point of view, hydroelectric is classified at the top of the list, and then wind energy, photovoltaic, and geothermal are lowest efficient renewable energy resources. Because of availability of cells in different categories, the efficiency of photovoltaic is very much variable.

Renewable Energy Resources

Characteristics of a renewable energy sources:

- Renewable energy sources are regenerated in natural processes.
- It can be used indefinitely.
- These are available in great amount or abundant in nature.
- It develops in a relatively short period of time.

These include solar energy; wind energy; ocean thermal power; wave and tidal energy; geothermal energy; hydropower etc. They are capable of solving the twin problems of energy supply in decentralized manner and helping in sustaining cleaner environment.

World Status of Renewable Energy Consumption

Renewable energy supplies 18% of the world's total energy consumption from traditional biomass, large hydropower and "new" renewables (small hydro, biomass, wind, solar, geothermal, and biofuels). Traditional biomass and hydropower represents about 71% and 16% respectively of total renewable energy. In terms of total energy share biomass and large hydropower represents 13 % and 3% respectively.

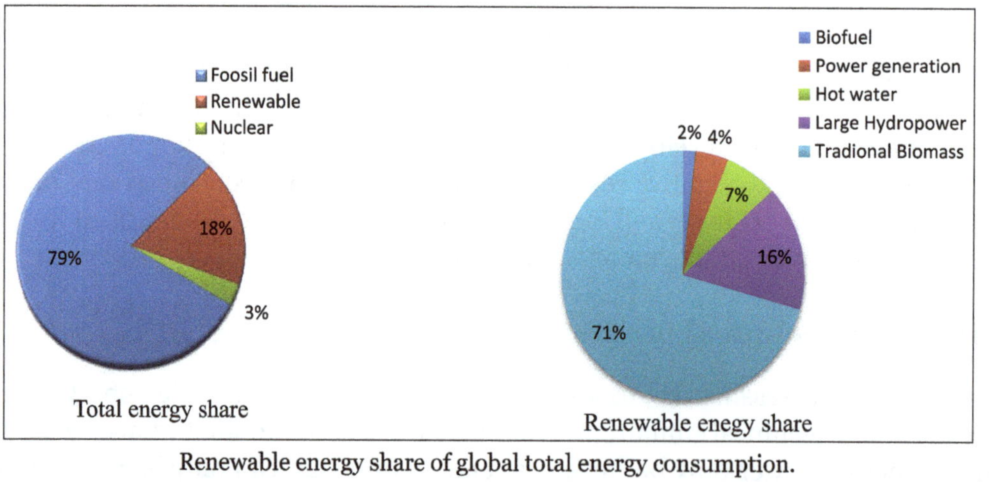

Renewable energy share of global total energy consumption.

Advantages of Renewable Energy

Renewable Energy is Eco-friendly

Renewable energy is considered clean energy since it doesn't cause grave environmental pollution, and it has low or zero carbon and greenhouse emission. Fossil fuels emit high levels of greenhouse gas and carbon dioxide, which are greatly responsible for global warming, climate change, and degradation of air quality. Fossil fuels also contribute to sulfur emission to the atmosphere leading to acid rains. Acid rains can cause damage to buildings.

Solar and wind power are considered eco-friendly because they emit zero toxic gases to the environment. The use of renewable energy dramatically reduces the dependence on fossil fuel as a source of energy, hence, cutting back on air pollution.

It's a Renewable Resource

This implies that they do not deplete over a lifetime, and there is zero possibility that they will run out (sustainable source of energy). Sources of energy like fossil fuels (oil, gas, and coal) are considered limited resources, and there is a strong possibility that they will run out in the future.

Renewable energy can help developing countries from over-reliance on fossil fuels. Powerful winds, heat emanating from beneath the earth, sunshine and moving water can guarantee a huge and steady energy supply to a nation for many years.

Renewable Energy is a Reliable Source of Energy

In the previous few decades, the use of fossil fuel has sharply increased. This over-reliance on fossil fuels has led to our security being threatened. Fossil fuels are prone to trade disputes, political instabilities, a spike in energy prices and unnecessary wars. These variables affect a lot more than a nation's energy policies; they can significantly drain a county's economy.

Although most argue that solar and wind energy is unreliable, a solid infrastructure puts this argument to rest. If solar and wind plants are distributed over a large geographical location, there can be minimal electricity generation interruption because weather disruptions in one location cannot be the same in other locations.

Leads to Job Creation

Renewable energy makes real economic sense because it is a cheaper alternative to most traditional sources of energy. Since the inception of renewable energy, new and stable jobs have been added to most world economies. For instance, in Germany and UK, many jobs have already been created. Thanks to their relentless efforts to develop and encourage the use of renewable forms of energy. Experts project that with the ongoing rigorous campaigns to embrace renewable energy, thousands of stable jobs will be created.

Renewable Energy has Stabilized Global Energy Prices

Change up to renewable sources of energy means the stability of energy prices across the globe.

This is because the cost of renewable energy depends on the initial cost of installation of renewable energy technologies as opposed to fossil fuels, which increase and decrease depending on the current inflation and availability of the resource. Respective governments would only need to cater to the initial costs, and that's it.

Less Maintenance of Facilities

Renewable energy technologies require less overall maintenance than traditional generators that use traditional fuel sources. This renewable energy generating technology like solar panels and wind turbines either have few or no moving parts. Above all, they don't rely on flammable, combustible fuel sources to operate, which makes the operating costs lower too.

Moreover, once the infrastructure for the harnessing of the renewable resource is laid down, there is low to zero maintenance required. This means that the owners of the facilities will reap big profits while providing cheap electricity to the population.

Boosts Public Health

This is a no-brainer. The sources for renewable energy generation emit little to no greenhouse gases or pollutants into the air. This means a smaller carbon footprint and an overall healthier atmosphere.

If governments took upon themselves to build more renewable energy facilities, the population would enjoy the health benefits. According to a study by the U.S. Environmental Protection Agency, Americans spend approximately $361.7 to $886.5 billion every year on the overall health of the population.

A big chunk of this budget goes to mitigate and cure diseases related to fossil fuel use like respiratory and cardiac health issues, cancer, and neurological disorders. Greenhouse, carbon and sulfur compounds emitted by the combustion of fossils fuels are risky to our health if inhaled over time. This is reason enough to consider renewable energy moving forward.

Empowering of People in the Countryside

Renewable energy generation mainly takes place in remote settings. This means that local towns would get a fair share of power generated, ultimately, catalyzing the regeneration of those depressed areas both socially and economically.

Electrification of those areas will open up untapped opportunities for development through the advancement of greenhouses using geothermal power, district heating of towns and communities through hot water generated by the energy exploitation of forestry and agricultural biomass.

It is a Technology

Renewable energy technology is a term used to refer to hybrid technology or other forms of technology that store up energy created or produced by renewable sources. It can anticipate renewable energy supply and helps to deliver energy produced by renewable energy technologies to the final consumer.

Renewable energy technologies include wind power, solar power, hydroelectricity, biomass energy, and biofuels, which are chiefly used for transportation. Over time, technology becomes cheaper when production and installation methods are refined; therefore, renewables are on a path that will constantly become cheaper.

Renewables Save Money and Profitable

Renewable energy helps in saving money in the long term. You will save not only on maintenance costs but on operating costs as well. As you're using a technology that generates power from the sun, wind, steam, or natural processes, you don't have to pay to refuel.

The amount of money one saves using renewable energy can vary depending on a number of factors, including the technology itself. Generally, transitioning to renewable energy means roughly hundreds to thousands of dollars in savings.

It can be used to Recycle Our Waste Products

Biomass energy tends to have this specific benefit that more than any other form of renewables. Biomass consumes used organic products such as used vegetable oil, corn and soybean byproducts, or even algae to generate energy. At the same time, it reduces the amount of waste that goes into landfills, also reducing the amount of overall carbon that goes into the atmosphere.

Lower Reliance on Foreign Energy Sources

Renewable energy technologies allow people to produce energy locally. The more you're using renewable energy for your power needs, the less you'll be dependent on imported energy, and the more you'll contribute to the country's energy independence as a whole.

It is Offered in Multiple Formats

We use many different forms of renewable energy today, even without realizing it. Energy diversity is possible through renewable energy. One of the most common forms of renewable energy is hydropower. Dams are built on rivers, and the moving water runs turbines that are able to generate electricity.

New technologies in solar panels, such as road-based, could even allow for energy collection within current infrastructure layouts. There is also wind power, biomass, geothermal and ocean. That means a greater potential for power generation now and into the future.

References

- Renewable-energy-clean-facts: nrdc.org, Retrieved 04, April 2020
- Renewable-energy, science: britannica.com, Retrieved 22, July 2020
- Social-economic-and-environmental-impacts-of-renewable-energy-resources, wind-solar-hybrid-renewable-energy-system: intechopen.com, Retrieved 15, March 2020

2

Wind as a Source of Energy

Wind energy is the process of harnessing the power of the flowing wind or airflows that occur naturally in the Earth's atmosphere to create electricity. The simplest and most efficient method is a wind turbine. A wind turbine is a mechanically powered wind device that turns electrical generators for electrical power output. This chapter covers all aspects related to wind energy and its utilization.

The wind power is one of the indirect solar energy technologies. The wind is the air in motion resulting from the pressure gradient caused by solar radiation. About two per cent of the solar radiation reaching the earth's surface is converted to kinetic energy and this is sufficient to meet the energy demands if it is harnessed. The kinetic energy of the wind is utilized directly or converted to mechanical energy or used for electricity generation. Apart from its use for grinding grains and pumping water by wind mills, wind turbines are familiar for electricity generation. The nature of the terrain, degree of cloud cover and angle of the sun plays important role in wind generation of an area. Various factors affect the distribution of wind energy are as follows:

- The chain of mountains channelizes the air currents.
- The hills, trees and building act as obstructions and change the direction of airflow.
- The frictional effect of the surface determines the wind speed. This is the reason why wind speed is quite high at seashore as frictional effect is less on smooth surface or sea surface.
- Climatic disturbances resulting from rains affect wind speed.
- Topography of an area also affects wind speed, for example, wind can speed up when passing through narrow gap such as mountains gap.

Energy Available in the Wind

The power in the wind is computed by the concept of kinetics. The wind mill works on the principle of converting kinetic energy of the wind into mechanical energy.

Power = Energy/unit time

Energy available = Kinetic energy of the wind

Kinetic energy of any particle = $\frac{1}{2} mV^2$

Amount of air passing in unit time through an area A, with velocity V = A. V

Its mass is equal to its volume multiple by its density ρ of air,

$$m = \rho A.V$$

Where m is the mass of transversing the area A swept by the rotating blades of a wind mill type generator Substituting this value of mass ($m = \rho A.V$) in kinetic energy equation ($\frac{1}{2}mV^2$),

$$\text{Kinetic energy} = 1/2\rho A.V.V^2$$

$$= 1/2\rho A.V^3 \text{ watts}$$

The wind power available is proportional to air density. It is 1.225 kg/m^3 at sea level. The variation in air density over the year is 10-15 per cent due to changes in pressure and temperature.

The power is also proportional to intercept area. So an aeroturbine with large swept area has higher power than that with smaller area.

Since the area is normally circular of diameter D in horizontal axis aeroturbines,

$$A = \frac{\pi D^2}{4} \text{ sq.m}$$

Substitute $\left(A = \frac{\pi D^2}{4} \text{ sq.m}\right)$ in ($1/2\rho A.V^3$ watts), it becomes,

$$\text{Available wind power} = \tfrac{1}{2}\, \rho \frac{\pi D^2}{4}.V^3 \text{ watts}$$

$$= 1/8\rho.\pi D^2 V^3$$

Wind Energy Conversion System

The air circulation results from non- uniform heating of earth's surface by the sun. The expansion of air occurs immediately above a warm area which is forced upwards by cool, denser air which flows in from the adjacent area. The energy in wind is due to its motion.

The wind energy is harnessed by wind turbine which is mounted on a tower. The speed of the wind depends directly with height. The tower height equal or more than 30 meters (100 feet) is beneficial as it can use faster and less turbulent wind. The wind speed at a height of 10 m is 20-25 percent higher than the surface and 30-60 percent at a height of 60 m. This is due to reduction in drag effect of the earth's surface. The average wind speeds in a particular location need to exceed at least 6–8 metres per second (m/s) for a small wind turbine to be economically viable.

Wind energy conversion systems are referred as WECS or aerogenerators or wind turbine generators or wind turbines. The factors influence the output from a wind energy converter:

- Wind speed.
- Cross section of wind swept by rotor.
- Overall conversion efficiency of the rotor, transmission system and generator or pump.

The maximum wind energy available is proportional to the cube of the wind speed. So a small increase in wind speed may increase the wind power.

Lift and Drag Force

The basis for wind energy conversion is Lift and Drag. The extraction of power from the wind depends on creating certain forces and applying them to rotate a mechanism. There are two primary mechanisms for producing forces from the wind:

- Lift,
- Drag.

The lift forces act perpendicular to the air flow while drag forces act in the direction of flow. These forces are produced by changing the velocity of the air stream flowing over either side of the lifting surface speeding up the air flow causes the pressure to drop, while slowing the air stream down leads to increase in pressure. The change in velocity leads to pressure difference across the lifting surface. This pressure difference produce a force that begins to act on high pressure side and moves towards the low pressure side of the lifting surface is called airfoil.

A good airfoil has a high lift/drag ratio. The lift increases as the angle formed at the junction of airfoil and the air-stream becomes less acute upto the point where the angle of the air flow on the low pressure side becomes excessive. Due to turbulences the lift decreases and drag force increases substantially, which is known as stalling. For efficient functioning of the wind turbine there should be high lift but less drag force.

Magnus effect is that caused by spinning a cylinder inn an air stream at a high temperature. The spinning slows down the air speed on the side where the cylinder is moving into wind and increases it on the other side. The air is blown through narrow slots in a cylinder and it emerges tangentially is known as Thwaits slot. This creates rotation of the air flow which in turn generates lift.

Different Wind Velocities

- Wind velocity (μ_0). The velocity of free air stream at sufficient distance away from wind turbine where there is no disturbance due to rotation of the wind turbine.
- Incident wind velocity (μ_1).It is the velocity at which the wind strikes the blade this velocity is always slightly lower than the free air stream velocity (μ_0).
- Blade linear velocity (v).It is the velocity of the blade due to the rotation of blade. If ω is the angular velocity of the blade and R is the length of the blade, then blade linear velocity $\upsilon = \omega \times R$.
- Relative velocity(vr). It is the relative velocity of wind with respect to the moving blade. It is the vector sum of incident wind velocity(μ_1) and blade linear velocity (υ) mathematically, it can be given as:

  $$\vec{v_r} = \vec{\mu_1} + \vec{v}.$$

The specified wind speed at which a wind turbine's rated power is achieved is known as rated wind speed.

Survival wind speed/extreme wind speed: It is the maximum wind speed that a wind turbine is designed to withstand.

Angle of Attack or Angle of Incidence (α)

It is the angle between the centerline of the aerofoil (blade cross- section and the relative wind velocity μr). the airflow remain attached to the aerofoil for small angle of attack. The airflow is separated from the aerofoil for large angle of attack.

Pitch Angle or Blade Setting Angle (γ)

It is the angle between the centerline of the aerofoil and the direction of linear motion of the blade. The output of a turbine is greatly influenced by blade pitch angle. The blade pitch control is a very effective way to control the output power, speed or torque of the turbine.

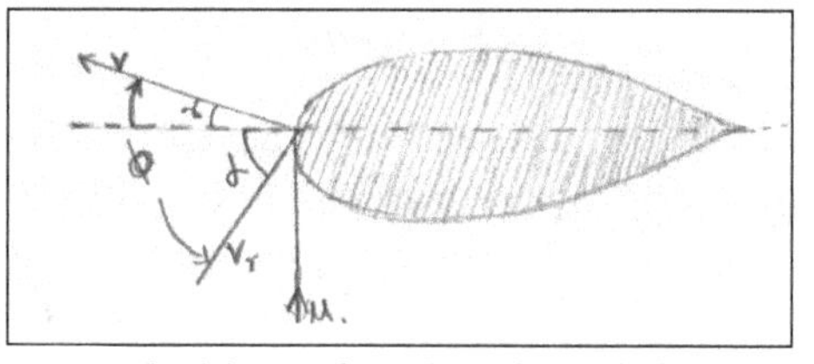

Velocities and angles of aerofoils.

Drag Force (ΔFD)

When a body (aerofoil) is placed in uniform airflow, there are two force acting on the body, namely, pressure force acting normally to the surface of the body, and shear force can be now resolved in the selection parallel to airflow is called drag force. The drag force always opposes the relative motion between the body and the air. It is given by the following equation:

$$\Delta F_D = C_D \times 1/2\ \rho \mu_1^2 .A$$

Where C_D is dragging coefficient and A is projected area of the body aerofoil perpendicular to the direction of air flow.

Lift Force (ΔFL)

The component of the total force (pressure force and shear force) on the body in the direction perpendicular to airflow is called lift force. As the name suggests, this force tries to lift the body. It is given by the following:

$$\Delta F_L = C_D \times 1/2\ \rho \mu_1^2 .A$$

Where, CL is lift coefficient.

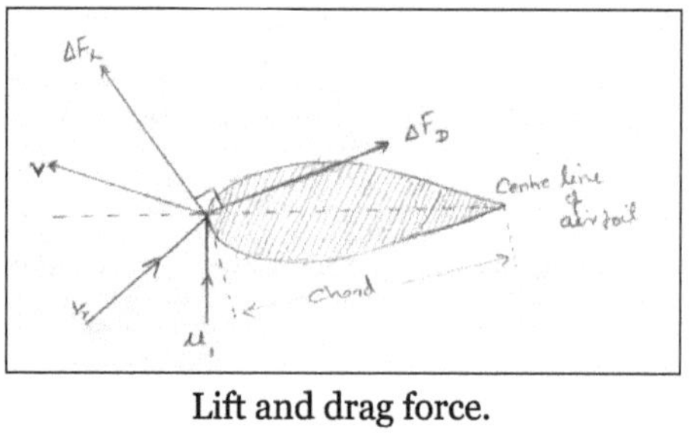

Lift and drag force.

Axial Force (ΔFA)

The total force (pressure force and shear force) can also be resolved along the axis of rotation of blade and perpendicular to it (tangential force on the blade). The component of total force acting on the blade along the axis of rotation of the blade is called the axial force. The axial force does not contribute to the rotation of the blade. It is also called thrust force and has to be balanced by a suitable reaction force generated by any thrust bearing provided on a rotor. The axial force contributes to waste energy which can be given as follows:

$$\Delta F_A = \Delta F_L \cos\phi + \sin\phi$$

$$\phi = \alpha + \gamma$$

Tangential Force (ΔFT)

It is the component of total force on the blade acting tangential to its circular path of rotation this is the force which contributes mainly to the useful energy extracted from the wind energy. It should be as high as possible. It is given by:

$$\Delta F_T = \Delta F_L \sin\phi + \Delta F_D \cos\phi$$

Parameters used in Rotor Design of Wind Turbine

Solidity

It is the ratio of projected area of blades to the swept area,

$$\text{Solidity} = \frac{\text{Projected area of blades}}{\text{Swept area (Example)}}$$

Example: A wind turbine has a rotor of 6m with 30 blades. Each blade has width of 0.30 m. Find the solidity?

$$\text{Solution: Solidity} = \frac{\text{Projected area of blades}}{\text{Swept area}}$$

$$= \frac{30 \times 0.30}{\pi x 6^2} = 7.96\%$$

The greater is the solidity of a rotor, the slower it needs to turn to intercept the wind. When wind will has a lesser number of blades, it needs to rotate fast to intercept the wind so that wind should not be lost through the large gap existing between two blade without importing a part of wind power.

Tip Speed Ratio (λ)

The tip speed ratio is defined as the ratio of the speed of tip of the rotor blade to the speed of coming air. Hence, tip speed ratio, TR is:

$$T_R\lambda = \text{Tip speed of blade/wind speed}$$

$$= \frac{\omega \times R}{\mu_o}$$

Where,

- ω is rotational speed of the rotor (radians/sec).
- R is radius of the rotor (meter).
- μ_o is wind speed or velocity (meter/second).

For a particular wind speed, there exists an optimum turbine tip speed to produce the maximum output. It is important to match the rotation of wind turbine to the corresponding wind speed.

The rotor blades of a wind turbine interact with wind. The slow rotation of the blades allow wind to pass freely through rotor swept area without interaction while for the fast movement rotor swept area acts as barrier deflecting win and cause turbulence thereby losing wind power.

The optimal tip speed ratio for horizontal-axis, three blade turbines, which are predominantly used in onshore and offshore wind farms, is approximately 5:1 to 6:1.

Example: Find the tip speed ratio if a 6 m diameter rotor has rotation of 20 rpm and the wind speed is 4 m/s? Solution: Tip speed ratio,

$$\lambda = \frac{\omega \times R}{\mu_o}$$

$$= \frac{\pi RN}{\mu_o}$$

$$= \frac{60}{4} = \frac{3.14 \times 6 \times 20}{60 \times 4} = 1.6$$

The wind mill rotating fast has tip speed ratio greater than one. Two or three blade rotors rotate faster and have tip speed ratio of 3 to 10. More bladed rotors rotate slowly having tip speed ratio between 1 and 2.

Principle of Power Generation from Wind

Wind turbine is used to extract useful energy from wind. The energy can be extracted by partially decelerating and expanding the airstream (reduction of pressure) using wind turbine. The rotor of the wind turbine collected wind from the whole area swept by the rotor. The area swept can be considered as airstream tube which is continuously expanding air. This airstream model is also called Betz model of expanding air. As air mass flow rate should be the same must decrease within the stream tube according to the law of continuity, the wind speed must decrease as air expands. As shown, airstream tube has area of A_0 at upstream, area of A_1 while passing through rotor blade (aerofoil) and area A_2 downstream.

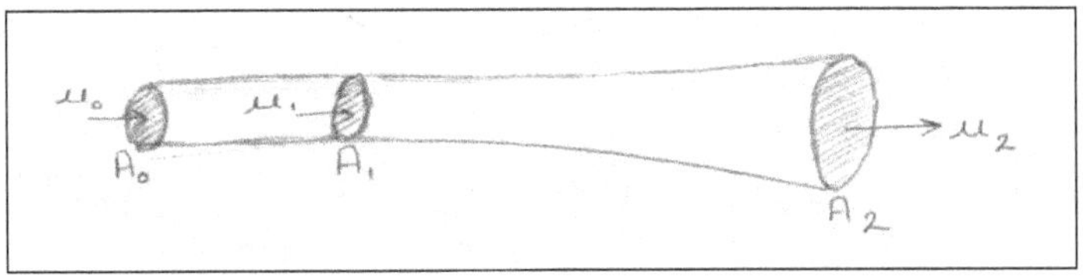

Airstream tube.

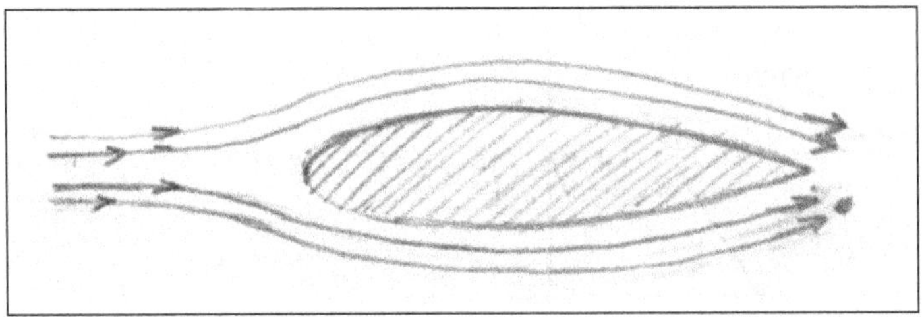

Airstream on aerofoil.

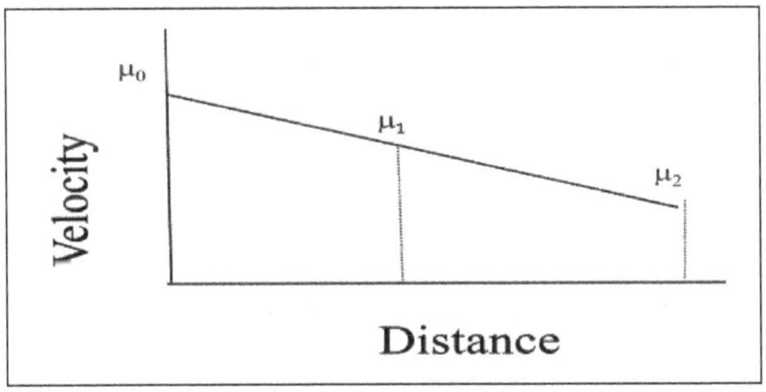

Variation of wind velocity.

Consider that μo and μ2 are wind velocities at upstream and downstream. The velocity reduction from μo to μ2 means that there is reduction in momentum of wind as it passes through wind turbine, resulting in a force being exerted on the blade of rotor, given by following equation,

$$\text{Force, } F = m \times \mu_o - m \times \mu_2 = m\left(\mu_o - \mu_2\right)$$

Where, 'm' is mass flow per unit time through stream tube.

The above force is exerted at a uniform rate when airflow moves over the rotor blade with velocity of μ1. Therefore power extracted is equal to the work done by airstream in moving for a distance of μ1 against force of F which is given by,

$$\text{Power of turbine, } P_T = F \times \mu_1$$

Substituting Equation $F = m \times \mu o - m \times \mu 2 = m\left(\mu_o - \mu_2\right)$ in $P_T = F \times \mu_1$,

$$P_T = m\left(\mu_o - \mu_2\right) \times \mu_1$$

The power extracted by the wind turbine from the wind is equal to the change or loss of kinetic energy of the wind.

$$\Delta KE = P_W = 1/2m\left(\mu_o^2 - \mu_2^2\right)$$

The energy extracted equal to energy lost by wind,

$$P_T = P_W$$

$$m(\mu_o - \mu_2) \times \mu_1 = 1/2m(\mu_o^2 - \mu_2^2)$$

$$\mu_1 = (\mu_o + \mu_2)/2$$

Where, μ1 is the average of μo and μ2.

Substituting Equation $\mu_1 = (\mu_o + \mu_2)/2$ in $P_T = m(\mu_o - \mu_2) \times \mu_1$,

$$P_T = m(\mu_o - \mu_2) \times (\mu_o - \mu_2)/2$$

$$= 1/2m(\mu_o^2 - \mu_2^2)$$

The mass flow (m) is the product of area, velocity and density,

$$m = A_1 \times \mu_1 \times \rho$$

$$= A_1 \times (\mu_o + \mu_2)/2 \times \rho$$

Substituting Equation $P_T = 1/2m(\mu_o^2 - \mu_2^2)$ and $= A_1 \times (\mu_o + \mu_2)/2 \times \rho$,

$$P_T = ¼A_1 \times \rho \times (\mu_o^2 - \mu_2^2) \times (\mu_o + \mu_2)$$

If an interference factor 'α' is there, then,

$$\mu_1 = (1-\alpha) \times \mu_0$$

$$\alpha = 1 - \mu_1/\mu_0 = 1 - (\mu_o + \mu_2)/2 \times /\mu_0 = (\mu_{0-}/\mu_2)/2_{\times}\mu_0$$

$$\mu_2 = (1-2\alpha) \times \mu_0$$

Then turbine power is:

$$P_T = ¼\ A_1 \times \rho \times (2 \times \alpha \times \mu_0) \times [\mu_o + (1-2\alpha)\mu_0]^2$$

$$= 4 \times \alpha \times (1-\alpha)^2 [1/2\rho \times A_1 \times \mu_o^3]$$

And Wind power is given by,

$$P_W = 1/2\ \rho \times A_1\ \times \mu_o^3 \text{ substituting 'm'} = A_1 \times (\mu_o + \mu_2)/2 \times \rho$$

From Equation $P_T = ¼A_1 \times \rho \times (\mu_o^2 - \mu_2^2) \times (\mu_o + \mu_2)$ and $P_T = 4 \times \alpha \times (1-\alpha)^2 [1/2\rho \times A_1 \times \mu_o^3]$, we have,

$$P_T = 4 \times \alpha \times (1-\alpha)^2 \times P_W$$

$$=C_p \times P_W$$

Where, Cp is the power coefficient and is given by,

$$C_P = 4\alpha(1-\alpha)^2$$

The power coefficient, CP indicates the portion or fraction of wind power that is extracted by wind turbine. The ideal or maximum, theoretical efficiency,ηmax of a wind turbine, also known as power co-efficient is the ratio of maximum power obtained from the wind to the total power available in the wind. It is the ratio of actual power to the theoretical power.

$$\text{Power co-efficient, } C_p = \frac{\text{Power output from wind machine}}{\text{Power available in wind}}$$

$$= P_{actual} / P_{theoretical}$$

- When $\alpha = 0$, then $\mu_1 = \mu_0, \mu_2 = \mu_0$ and no power generation takes place.
- When $\alpha = 1/3$, then $\mu_1 =_{2/3} \mu_0, \mu_2 = 1/3\mu_0$ and maximum power generation takes place.
- When $\alpha = 1/2$, then $\mu_1 =_{1/2} \mu_0, \mu_2 = 0$ and only turbulence occurs at downstream.
- When $\alpha = 1$, then μ_1 and it results in stalling of turbine.

Theoretical Maximum Wind Power

The air density is proportional to the air temperature and the air pressure, both of which vary with height above sea level. The power in the wind cannot be completely converted to mechanical energy of a wind turbine. The losses due to friction reduce the actual wind power. The theoretical maximum of energy extraction from wind was calculated by Betz in 1926. The power output from wind turbine is given as follows:

$$P_T = \tfrac{1}{4} A_1 \times \rho \times \left(\mu_o^2 - \mu_2^2\right) \times (\mu_o + \mu_2)$$

The power output depends on outlet velocity μ2, Hence for the maximum output dPT/dμ2 =0 and so μ2 = -μ1, which is impossible μ2 = μ1/3.

Substituting μ2 = μ1/3 in Equation $P_T = \tfrac{1}{4} A_1 \times \rho \times \left(\mu_o^2 - \mu_2^2\right) \times (\mu_o + \mu_2)$, then

$$P_T = 8/27\ \rho \times A \times \mu_0^3 = 16/27\left[1/2 \times \rho \times A \times \mu_0^3\right]$$

$$= 16/27 \times P_W$$

$$= 59.3\% \times P_W$$

As per Betz theory a wind turbine able to utilize only 59 percent of the wind power even if no losses has occurred and so the power is multiplied by a factor of 0.59. When some unavoidable swirl losses occur the turbine is able to utilize only 42 percent and the value reduces to 0.42.

Almost all large wind turbines are of horizontal axis wind turbine designs having an upwind rotor with three blades, attached to a nacelle on top of a tall tubular tower. So the wind energy is captured by the rotor blades in the turbine. So the kinetic energy is converted by wind turbines to mechanical power. The conversion of kinetic energy into mechanical or electric energy is:

$$\text{Pmech} = M\Omega = M 2\pi n$$

Where,

- M - torque,
- n - rotational number,
- Ω - angular velocity.

The mechanical power in the rotation of blades is as follows:

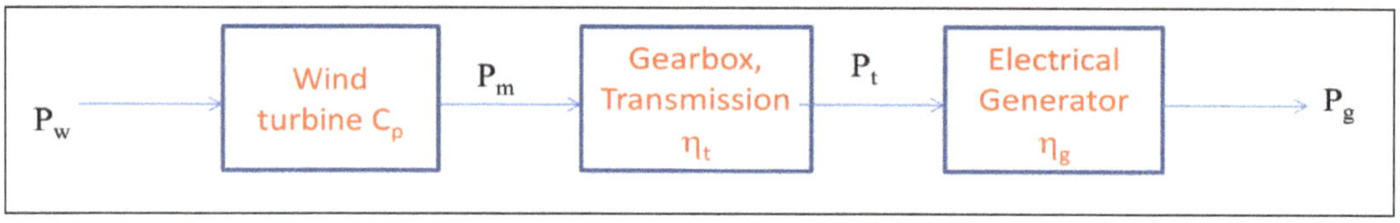

Wind turbine efficiencies.

The mechanical power in the rotor blades is coupled to a transmission or gear box having the transmission or gearbox efficiency.

The optimization of wind turbine designs to suit the different wind conditions and use of a variable pitch control system, yaw control, large sized blades, variable speed drive, tilt control of blades, stall control by shifting of blades, eddy current bracking system to maintain speed, power electronics can increase the efficiency of a wind turbine.

Selection Criteria for Wind Turbine Siting

The main consideration for selecting a site for wind generator is as follows:

- High annual mean wind speed: A basic requirement for a successful use of a windmill of farm is an adequate supply of wind and good wind speeds. The wind power is proportional to the cubic power of wind speed.
- No obstruction: There should not be any high structure to obstruct wind for a distance of 3 km to the windmill.
- Open plain: The site should be the open plain such as open sea shoreline when the windmill is located on hill or a ridge with gentle slope.
- Height: The wind speed increase with height, which can be obtained when the windmill is located on a hill or a ridge with gentle slope.
- Near lake or ocean: Differential heating of water and land generates wind of sufficient speeds.
- Topography: Topography such as mountain gap helps to channelize and speed up winds.
- Favourable land cost: It helps in restriction or reducing the cost of project.

- Nearness to load centre: It reduces the cost of transmission of the generated power.
- Nearness to road or rail link: It helps in installation of windmill.
- Availability of wind rose: It helps in designing of windmill as wind data of the site can be determined.

Application of Wind Energy

There are two types of small scale wind turbines utilized in domestic, community and wind energy projects. The power generated by wind turbines are connected to grid (state or central). The first windmill used for the production of electric power was built in Scotland in July 1887 by Prof James Blyth of Anderson's College.

- Stand-alone systems: Stand-alone systems are used for electricity production for battery charging. It is used for running small electrical appliances mainly in remote areas with no power supply. It helps in water heating or pumping, electric livestock fencing, lighting or any kind of small electronic system needed to control or monitor remote equipment.
- Grid-connected Small Wind Turbines: The output from the wind turbine is directly connected to the existing mains electricity supply. This type of system can be used both for individual wind turbines and for wind farms exporting electricity to the electricity network.

Wind Farms

The group of wind turbines which are located in the same place for the electricity generation is known as wind farm or wind power plants. It varies in its capacity of few Megawatts to hundreds of Megawatts. A large wind farm may consist of several hundred individual wind turbines distributed over an extended area, but the land between the turbines may be used for agricultural or other purposes. Gansu Wind Farm is the largest wind farm in the world. A wind farm may also be located offshore.

Advantages of Wind Energy

Wind Energy is a Clean Source of Power

The production of wind energy is "clean." Unlike using coal or oil, creating energy from the wind doesn't pollute the air or require any destructive chemicals. As a result, wind energy lessens our reliance on fossil fuels from outside nations as well, which boosts our national economy and offers a variety of other benefits as well.

Renewable Source

The wind is free. In the event that you live in a geological area that gets a lot of wind, it is ready and waiting. As a renewable asset, wind can never be drained like other regular, non-renewable assets.

The expense of delivering wind energy has dropped fundamentally lately, and as it becomes more popular with the general population, it will just continue to be cheaper. You will recover the expense of obtaining and introducing your wind turbine over time.

Winds are caused by rotation of the earth, heating of the atmosphere by the sun, and earth's surface irregularities. We can harness wind energy and use it to generate power as long as the sun shines and the wind blows.

Wind Energy has Low Operating Costs

The wind farms or individual turbines can be expensive to install. However, once it is installed and running, operating costs are relatively low; fuel (wind) is free, and the turbines don't require too much maintenance over the course of their life.

Cost-Effective

Wind turbines can give energy to numerous homes. You don't actually have to possess a wind turbine, keeping in mind the end goal to harvest the profits; you can buy your power from a service organization that offers wind energy for a specific area. That means you don't even necessarily have to invest any cash in order to reap the benefits of wind energy for your home or business.

Prices are Decreasing

Prices have decreased by over 80% since 1980. Thanks to technological advancements and increased demand, prices are expected to keep decreasing in the foreseeable future.

Extra Savings for Land Owners

Landholders who rent area to wind homesteads can make a considerable amount of additional cash, and wind energy likewise makes new employments in this developing engineering field. Government organizations will also pay you if they can install wind turbines on your land. Also, in some cases, the electric company may wind up owing to you.

Use of Modern Technology

Wind turbines are considered by some to be incredibly attractive. The newest models don't look like the clunky, rustic windmills of old times. Instead, they are white, slick, and modern looking. That way, you don't have to worry about them becoming an eyesore on your land.

The latest advances in technology have transformed preliminary wind turbine designs into extremely efficient energy harvesters. Turbines are available in a wide range of sizes for farms, factories, and large private residences, extending the market with many different types of businesses and by individuals for use at home on larger plots and other plots of land.

Portable wind turbines are also available and can power small devices on the go. The latest models will generate even more electricity, require less maintenance, and run more quietly and safely.

Wind Power has seen Rapid Growth

Wind energy has seen enormous growth in the last decade. Cumulative wind power capacity increased by an average of 30% per year. Wind energy accounts for about 2.5% of the total worldwide electricity production. Wind turbines are available in various sizes, which means a vast range of people and businesses can take advantage of it to produce power for their own use or sell it to the utility to reap some profits.

Huge Market Potential

The potential for wind power is huge. Several independent research teams have reached the same conclusions, and that is the worldwide potential of wind power is more than 400 TW (terawatts). Harnessing wind energy can be done almost anywhere.

Great Potential for Residential uses

The wind energy is especially appealing to the residential market. People are able to generate their own electricity with wind power in very much the same manner as people do with the best solar panels (photovoltaic).

Wind is an independent energy source, and it's great for powering homes. In addition to this, wind-powered homeowners also gain access to something called net metering.Net metering basically provides credit to electricity bills for any excess power generated in a given month. Homeowners actually get paid for extra energy production, and that can even protect them from blackouts as well as fluctuating energy prices.

Wind Farms can be Built on Existing Farms

Wind turbines are incredibly space-efficient and can be installed on existing farms or agricultural land in rural areas where it can be a source of earning for the farmers as wind plant owners make payments to farmers for the use of their land for electricity generation. It doesn't occupy much space, and farmers can continue to work on the land.

At present, less than 1.5% of contiguous U.S. land area is used by wind power plants. However, if all the plains and cattle land made available on the interior of the country, there's a lot of opportunity for expansion if landowners and government land managers are up for it.

Conserves and Keeps Water Clean

Turbines produce no particulate emissions that contribute to mercury contamination in our lakes and streams. Wind energy also conserves water resources. For producing the same amount of electricity, nuclear power take about 600 times more water than wind, and coal takes about 500 times more water than wind.

The Wind Energy Industry Creates Jobs

The wind energy industry has boomed since wind turbines became commercially viable. As a result of this, the industry has created jobs all over the world. Jobs now exist for the manufacturing, installation, maintenance of wind turbines, and there are even jobs in wind energy consulting.

The renewable energy industry employed over 10 million people worldwide in 2017. Of these jobs, 1.15 million were in the wind power industry. China leads the way in providing over 500,000 of these jobs. Germany is in second place with around 150,000 jobs, and the United States is a close third with around 100,000 wind energy jobs.

Disadvantages of Wind Energy

Wind Reliability

Wind doesn't generally blow reliably, and turbines usually function at about 30% capacity or so. In the event that the weather is not going to support you, you may wind up without power (or at any rate you'll need to depend on the electric company to take care of you during those times). Serious storms or high winds may cause harm to your wind turbine, particularly when they are struck by lightning.

Wind Turbines could be Threat to Wildlife

The edges of wind turbines can actually be unsafe for wildlife, especially birds and other flying creatures that may be in the area. There isn't really a way to prevent this, but it's definitely something that you want to make sure that you are aware of being possible consequences that may come up as a result of it.

Wind Turbines could Lead to Noise and Visual Pollution

Wind turbines can be a total and complete pain to install and deal with on a regular basis. Wind turbines make a sound that can be between 50 and 60 decibels, and if you have to put it next to your home. Some individuals believe that wind turbines are ugly, so your neighbors may also complain about them.

While most people like how wind turbines look, few people like them but with NIMBY ("not in my backyard") attitude, but for the rest, wind turbines remain unattractive as they have a concern that it may tarnish the beauty of landscapes.

Are Expensive to Set Up

The manufacturing and installation of wind turbines require heavy upfront investments both in commercial and residential applications. Wind systems can involve the transportation of large and heavy equipment, causing a large temporarily disturbed area near the turbines. Erosion is another potential environmental problem that can stem from construction projects.

Wind turbines and other supplies needed to make wind energy could be extremely costly in advance, and relying upon where you live, it might be hard to find someone to sell them to you and somebody who can maintain it over time.

Cost Trade-off

The cost-competitiveness of wind power is highly debatable. Both utility-scale wind farms and small residential wind turbines typically rely heavily on financial incentives. To give wind power a fair chance in the fierce competition against already well-established energy sources such as fossil fuels and coal, financial incentives are crucial.

Wind turbines make an excellent alternative in some situations for a homeowner who wants to become an energy producer, but it would require wind turbines about 10 kilowatts and $40,000 to $70,000 to become a net electricity producer. Investments like this typically break even after 10 to 20 years, which is a pretty long time.

Safety of People at Risk

Severe storms and high winds can cause damage to the blades of the wind turbines. The malfunctioned blade can be a safety hazard to the people working nearby. It may fall on them, causing life term physical disability or even death in certain cases.

Wind Power can be Harnessed at Certain Locations Only

Wind energy can only be harnessed at certain locations where the speed of the wind is high. Since they are mostly set up in remote areas, transmission lines have to be built to bring the power to the residential homes in the city, which requires extra investment to set up the infrastructure.

Shadow Flicker

Shadow flicker occurs when the blades of the rotor cast a shadow as they turn. Research has shown the worst-case conditions would affect, by way of light alteration, neighboring residents a total of 100 minutes per year, and only 20 minutes per year under normal circumstances. Designers of wind farms avoid placing turbines in locations where shadow flicker would be a problem for any significant amount of time.

Effect on the Environment

It obliges a ton of open area to set up wind turbines and chopping down trees kind of eliminates the whole green thing that you're trying to do with them. Places that may be good for it may be difficult to get to and use. Consistency with city codes and mandates may be irksome when you are attempting to install a wind turbine. Sometimes, height confinements may keep you from installing one on your property as well.

Wind Farms

Wind farms are areas where many large wind turbines have been grouped together. They "harvest" the power of the wind. These large turbines look a bit like super-tall windmills. A large wind farm can have hundreds of wind turbines spread out over hundreds of miles. The land between the turbines may be used for other purposes, such as regular farming. Some wind farms are also located near bodies of water. There, they take advantage of winds that blow across lakes or oceans.

Did you know that wind energy is actually another form of solar energy? Earth's shape and rotation work with the Sun's uneven heating of the atmosphere to make winds. Wind farms are built in areas known to be especially windy on a regular basis. The winds turn the blades of the turbines. Then, the turbines turn the energy of the wind into mechanical power. Generators then turn the mechanical power into electricity. That electricity is then used to power homes.

You can think of a wind turbine as the opposite of a fan. A fan uses electricity to make wind. Wind turbines do the opposite: they use the wind to make electricity! As the wind turns the blades of a wind turbine, the blades cause a shaft to spin. The spinning shaft connects to a generator that creates electricity.

Are you wondering why scientists looked to the wind as an energy source? There are plenty of good reasons. Wind energy is free and renewable. Unlike most power plants, wind farms don't emit pollution or greenhouse gases.

However, wind farms can cost a lot of money to set up. Over time, though, their cost is competitive with other types of generating systems. Unfortunately, you can't make the wind blow whenever you want it to. That means wind farms can't always meet electricity needs on demand.

Wind Turbines

A wind turbine is a system that converts the kinetic energy available in the wind into mechanical or electrical energy. Parts of a wind turbine system:

- Foundation,
- Tower,
- Nacelle,
- Hub,
- Rotor,
- Drive -train,
- Gearbox,
- Generator,
- Electronics & Controls,

- Yaw,
- Pitch,
- Braking,
- Cooling.

In general the parts of a wind turbine system are grouped into:

- Rotor: The two types of rotors are Horizontal axis rotor and Vertical axis rotor. The vertical axis machines operate in all wind directions and no need of yaw adjustment.
- Wind mill head or turbine consists of rotor, housing of the rotor bearings. It houses control mechanism incorporated like changing the pitch of the blades for safety devices and tail vane to orient the rotor to face the wind.
- Transmission and control: The rate of rotation of large wind turbine generators operating at rated capacity or below is controlled by varying pitch of rotor blades. The transmission options include mechanical systems involving fixed ratio gears, belts and chains singly or in combinations or hydraulic systems involving fluid pumps and motors. Fixed ratio gears are recommended for top mounted equipment.
- Supporting structures.

The figure gives the detailed diagram of different parts of wind turbine.

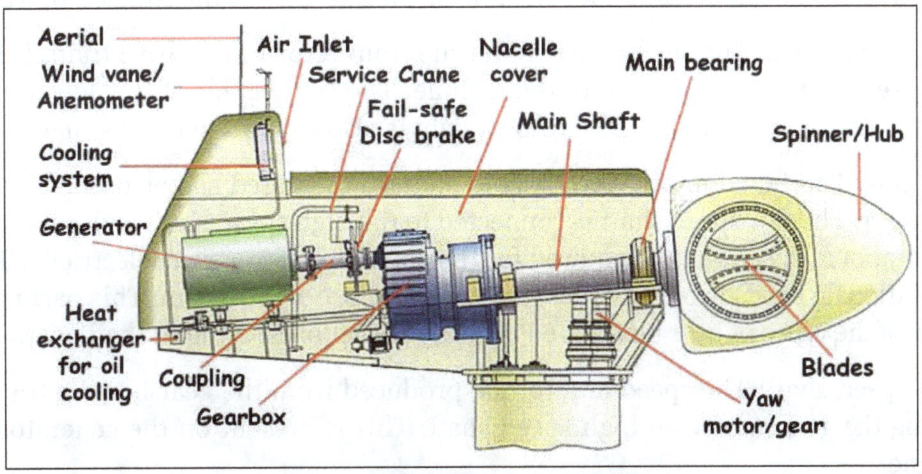

Diagram of different parts of wind turbine.

- Rotor: The component of wind turbine that is rotating and helps in conversion of kinetic energy to mechanical energy. It consists of blades and hubs.
- Hub: The function of rotor hub is to connect rotor blades to the rotor shaft. This component controls the power generation of the wind turbine.
- Blades: This is a rotating component of the system. This component is based on the principle of lift and drag (principle of aerodynamics). It converts kinetic energy first to mechanical energy and then transferred through shaft to generator for converting into electrical energy.

Two or three blades are common in the wind turbines. The wind blown over the blades lift the blades and rotate it. The two bladed wind turbines have lighter hub and so the whole structure is lighter. But three bladed wind turbines are aerodynamically efficient and have low noise.

The length of the blade is the important parameter for estimation of wind power generation potential of a wind turbine. The torque increases with more number of blades.

The blades are commonly made from composites mainly fiberglass or glass reinforced plastics (GRP) or carbon fiber reinforced plastics (CFRP). Wooden or epoxy laminates are also used for making blades.

Solidity is defined as the ratio of the projected area of the rotor blades on the rotor plane to the swept area of the rotor. High- solidity rotors use force for rotation and these rotor turn slower. Low- solidity rotors have slender aerofoil and these rotors use lift force for rotation and turn faster.

Chord: It is the width of the blade which is across distance from one edge of the blade to the other edge.

- Drive train: The drive train consists of low –speed shaft, gear box and a high speed shaft. It also has support bearings, couplings, brake and rotating parts of generator.
- Nacelle: The covered part of the wind turbine system over the top of tower is nacelle. It houses gear box, low speed shaft and high-speed shaft, generator, controller, and brake. It has an important role in protection of components of wind turbine from the various weather conditions. It also helps in reduction of noise produced from the rotation of wind turbine.
- Low-speed shaft: This shaft transfers torque from rotor to the drive train. It is a rotating part present on the rotor side of the turbine. The connection of this low speed shaft increase rotations per minute. The rotor weight is also supported by this shaft.
- Gear box: The function of gear box is to step up the speed as per needed by the electric generator. The low speed shaft is connected to the high speed shaft with gears. It increases the rotational speed that is required for the generator to generate electricity. The increase in rotational speeds is of the order 30-60 rpm to 1000-1800 rpm. This part is very costly. Some of the types of gear boxes are Planetary Gear Boxes, Parallel shaft gear.
- High-speed shaft: The speed and torque produced from the gear box are transmitted for driving the generator with high speed shaft. This is present on the generator side of the turbine.
- Brake: This part is meant to stop the running of wind turbines during extreme weather conditions. The various types of brakes are:
    - Mechanical brake (Disc brake, clutch brake).
    - Aerodynamic brake (Tip brake and spoilers).
- Generator: The conversion of rotational energy to electrical energy is carried out by generator. In general the wind driven electric generator produces 50-cycle AC electricity. The types of generators are:
    - Synchronous generator (Electrically excited, permanent magnet).

- Asynchronous generator (SQIG -Squirrel cage induction generators, Slip ring).

- Controller: The grid quality electric current is controlled by the controller of the turbine system. The controller starts up the machine at cut-in wind speed (generally 3 m/s) and shuts off the machine at cut-out wind speed (generally 25 m/s) as per the design requirement. The controller measures and controls parameters like Voltage, current, frequency, Temperature inside nacelle, Wind direction, Wind speed, The direction of yawing, shaft speed, Over-heating of the generator, Hydraulic pressure level, Correct valve function, Vibration level, Twisting of the power cable, Emergency brake circuit, Overheating of small electric motors for the yawing, hydraulic pumps, Brake caliper adjustment etc.

- Anemometer: Anemometer is an instrument that measures wind speed. It gives the input to the controller for power regulation and braking beyond the cut out & survival wind speed. Generally the anemometer is fixed on top of wind turbine.

- Pitch: The electricity production is controlled by pitch under different wind intensities. Blades are turned or pitched, out of the wind to control the rotor speed and keep the rotor from turning in winds that are too high or too low to produce electricity.

- Tower: This helps to use the wind energy at sufficient heights above ground. This helps to absorb and securely discharge static and dynamic stress exerted on the rotor, the power train and thc nacelle into the ground. The major types of towers used in wind turbine are:
    - Lattice tower,
    - Tubular tower,
    - Guyed tower,
    - Hybrid Tower.

The characteristics of different towers are given in table.

Tower type	Area of contact	Loading	Load distribution
Tubular steel tower	More	More	Even
Tubular concrete	More	More	Even
Lattice tower	Less	Less	Uneven
Three legged tower	Less	Less	Uneven
Guy wired tower	Less	Less	even
Hybrid tower	A combination of tubular and lattice		

The tubular towers are more popular among modern turbines because of their lower airflow interference and downstream turbulence creation apart from its more aesthetic acceptability. The tubular steel tower is costly. The transportation and fabrication of lattice tower, three legged tower and guy wired tower are easy. But guy wired tower is suitable for small wind turbines.

Generally tower height is 1 to 1.5 times the rotor diameter and it is normally at 20 m. The site characteristics have a great influence on tower selection. Tower stiffness is important factor in wind turbine system because of possibility of coupled vibrations between rotor and tower. Tower shadow is the wake created by air flow around a tower. There is an effect of tower shadow on turbine dynamics, power fluctuations and noise generation especially on downwind wind turbines.

- Foundation: The foundation type of foundation has been chosen based on soil conditions and water table location present at the planned site of a wind turbine:
    - Onshore Foundation Types: Slab Foundation (preferred when the top soil is strong).
    - Pile Foundation (Preferred when the top soil is of a softer quality).
    - Offshore Foundation Types: Monopile, Gravity base, Tripod.
- Wind vane: This measures the wind direction. It provides information to the controller for orienting the turbine properly (yawing) with respect to the wind direction.
- Yaw drive: The function of this is to keep the turbine aligned to the wind. It turns the nacelle with rotor as per the wind direction using a rotary actuator engaging on a gear ring beneath the nacelle. Two types of yaw drive systems are active yaw or free yaw systems. Active yaw drive mechanism is with yaw motors and is controlled by automatic yaw control system with its wind direction sensor mounted on nacelle in position when it is not in yawing. It is used commonly with upwind wind turbines. Free yaw systems can self-aligned with the wind and commonly used on downwind wind turbine systems.
- Yaw motor: The yaw drive is powered by the yaw moto.

Classification of Wind Turbines

There are different types of turbine technology:

- Based on axis orientation:
    - Horizontal,
    - Vertical.
- Based on Number of Blades:
    - One,
    - Two,
    - Three,
    - Multi-bladed.
- Based on Location:
    - On-shore,
    - Off-shore.
- Based on Power control:
    - Stall,

    - Variable Pitch,
    - Controllable Aerodynamic.
- Based on Surfaces/Yaw Control Yaw Orientation:
    - Driven Yaw,
    - Free Yaw,
    - Fixed Yaw.
- Based on Rotor Position:
    - Upwind,
    - Downwind.
- Based on Transmission:
    - With gear,
    - Without gear.
- Based on Type of Hub:
    - Rigid,
    - Teetered,
    - Hinged blades,
    - Gimbaled.
- Based on Generator Speed:
    - Constant,
    - Variable.

The common classification is based on axis type:

- Horizontal axis Wind turbine (HAWT),
- Vertical axis Wind turbine (VAWT).

Horizontal Axis Wind Turbine (HAWT)

The most common type of wind turbine is Horizontal axis Wind turbine (HAWT). The axis of the rotation of this wind turbine is parallel to the ground. In HAWT, the axis of blade rotation is also parallel to the wind flow. Commonly found horizontal axis wind mills are aero-turbine mill with 35% efficiency and farm mills with 15% efficiency. The main components in HAWT are:

- Rotor with blades for the conversion of wind energy to rotational energy. It costs 20% of the wind turbine cost.
- Generator component consists of electrical generator, the control systems and gearbox. It

converts low speed rotational energy to high speed rotational energy needed for electricity generation. It contributes to 34 % of wind turbine cost.

- Structural support includes tower and yaw mechanism and it costs 15% of the turbine cost.

The figure represents the main components of a HAWT.

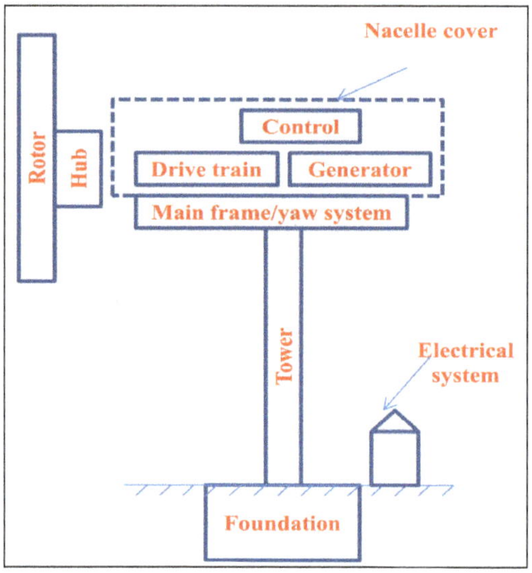

Main components of Horizontal axis wind turbine system.

The HAWT works on principles of aerodynamic lift of wind turbines.

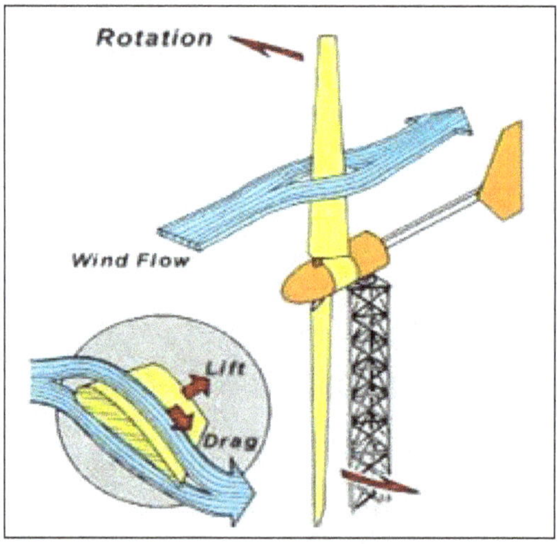

Principle of working of HAWT.

Classification of HAWT

Classifications of HAWT are based on:

- Rotor orientation:
    - Upwind of the tower,
    - Downwind of the tower.

- Hub design:
  - Rigid,
  - Teetering.
- Rotor control:
  - Pitch,
  - Stall.
- Based on number of blades:
  - Two blades,
  - Three blades.
- Based on how they are aligned with the wind:
  - Free yaw,
  - Active yaw.

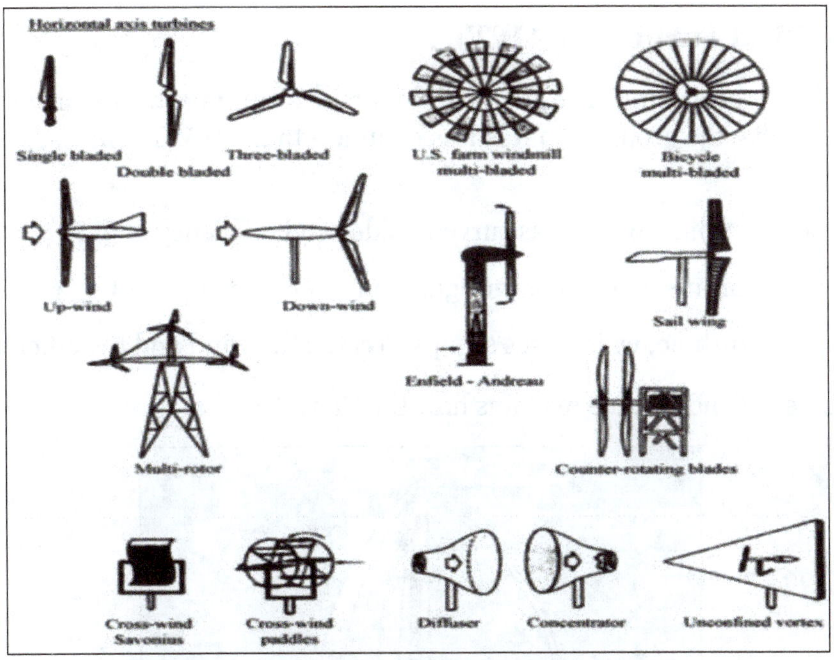

Various concepts for horizontal axis turbines.

Examples of HAWT are:

Single blade, two blade, three blade and Multi blade HAWT.

The advantages of HAWT are:

- More table due to position of blades of the HAWT to the side of the turbine's center of gravity.
- Efficiency is more because of maximizing collection of wind energy by turbine by adjusting angle of attack.
- During storm rotor blades are able to pitch and reduce the damage.
- Accessibility to strong wind is possible in HAWT (tall tower).
- It is suitable for areas with wind shear and placement on uneven land or in offshore locations. But it has disadvantages like.
- Facing difficulty in operating near the ground.
- Difficulty in transporting heavy and tall towers and long blades.
- Requires a special installation procedure.
- It creates navigation problem at offshore sites.

Vertical Axis Wind Turbine (VAWT)

The axis is parallel to the ground. The VAWT do not take advantage of the higher wind speeds at higher elevations above the ground. So it is less common than HAWT. The basic vertical axis designs are:

- Darrieus wind turbine, which has curved blades and efficiency of 35%.
- Giromill wind turbine, which has straight blades, and efficiency of 35%.
- Savonius wind turbine, which uses scoops to catch the wind and the efficiency of 30%.

The figure of Darrieus wind turbine explains how the VAWT works.

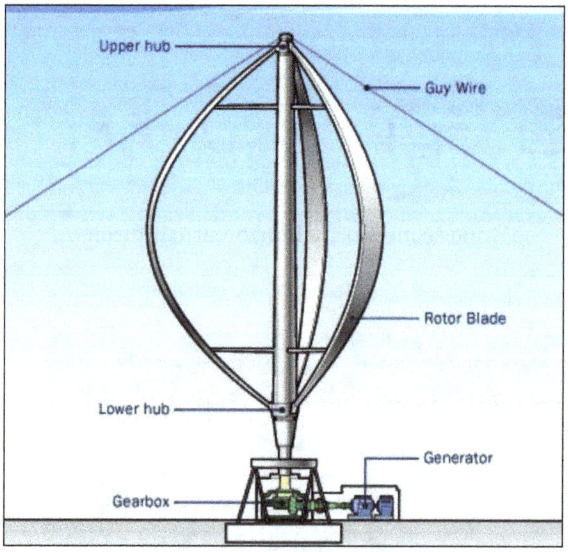

Darrieus wind turbine.

(a) Helical Giromill Wind Turbine.

(b) Giromill Wind Turbine.

(c) Savonius Wind Turbine.

The advantages of VAWT are:

- No need ofwind turbine to orient with respect to wind direction.
- Lighter weight.
- Easy maintenance because the shaft is vertical, the transmission and generator can be mounted at ground level.
- Lower cost tower.

But VAWT is not as efficient at collecting energy from the wind in comparison to HAWT.

Parameters Influencing Wind Energy Conversion

The wind energy conversion depends on:

- Optimum tower height.

- Control systems.
- Number of blades and blade shape.

Designing of Wind Turbine

The designs of wind turbine are based on:

- Number of blades.
- Rotor orientation.
- Blade material, construction method.
- Hub design.
- Power control.
- Rotor speed (fixed or variable).
- Orientation of turbine.
- Type of generator.
- Gear box or direct drive generator.

The different parameters of wind turbine like solidity, blade tip speed ratio, rotational speed are also important.

Solidity = Projected area of blades/swept area

Example: A wind mill has rotor of 6m with 30 blades. Each blade has width of 0.39. Calculate solidity?

Solution: Solidity = Projected area of blades/swept area $= \dfrac{30 \times 0.3}{\pi \times 6^2} = 7.96\%$

The blade tip speed can be calculated from the rotational speed of the turbine and the length of the blades used in the turbine using the following equation:

$$\text{Blade tip speed} = \frac{\text{rotational speed }(\text{rpm}) \times \pi \times \text{D}}{60}$$

Where, D is the diameter of the turbine.

The tip speed ratio is very important factor in the different formulas of blade design.

Tip Speed Ratio (TSR), λ = (tip speed of blade)/(wind speed).

For a particular wind speed, there exists an optimum turbine tip speed for producing maximum output. The slow running multi bladed wind turbine rotors operate with tip speed ratios in the range of 1-4 and fast runners use 5-7 as tip speed ratios. So it is important to match the rotation of wind turbine to the corresponding wind speed.

Example: Estimate λ if the rotational speed of the turbine is 20 rpm and diameter of 6m and wind speed of 4m/s?

Solution: Blade tip speed $= \dfrac{\text{rotational speed (rpm)} \times \pi \times D}{60}$

$$= \frac{20 \times \pi \times 6}{60} = 6.28 \text{m/s}$$

Tip Speed Ratio (TSR), λ = (tip speed of blade)/(wind speed)

= 6.28/4=1.57 ~ 1.6

How generator capacity, rotational speed, wind speed and swept rotor area of the turbine are related?

Two formulas are needed to explain it.

Power (W) = $0.6 \times Cp \times N \times A \times V^{3,}$

Rotational speed (rpm) = $V \times TSR \times 60/(6.28 \times R)$,

Where Cp = Rotor efficiency, Normally Cp = 0.4 is used in calculations:

- N = Efficiency of driven machinery,
- A = Swept rotor area (m^2),
- V = Wind speed (m/s),
- TSR = Tip Speed Ratio,
- R = Radius of rotor.

This concept works without transmission.

If a transmission with efficiency of 0.95 was to be included this means then,

$N = 0.95 \times 0.7$

How to Find Out Blade Chord?

Blade Chord (m) = $5.6 \times R^2/(i \times Cl \times r \times TSR^2)$,

Where,

- R = Radius at tip,
- r = radius at point of computation,
- i = number of blades,
- Cl = Lift coefficient,
- TSR= Tip Speed Ratio.

Example: A wind mill having tip speed ratio of 7, rotor efficiency of 0.4, generator efficiency of 0.7, swept rotor area of 2.11 m² and radius of rotor is 0.82 m. And the wind speed of 8.6 m/s. Calculate rotational speed and power output of the turbine?

Solution:

- Rotational speed = $V \times TSR \times 60/(6.28 \text{ x } R) = 8.6 \times 7 \times 60/(6.28 \times 0.82) = 701$ rpm.
- Power output = $0.6 \times Cp \times N \times A \times V^3 = 0.6 \times 0.4 \times 0.7 \times 2.11 \times (8.6)^3 = 226$ W.

The figure shows the wind turbine design and power output progression over the years.

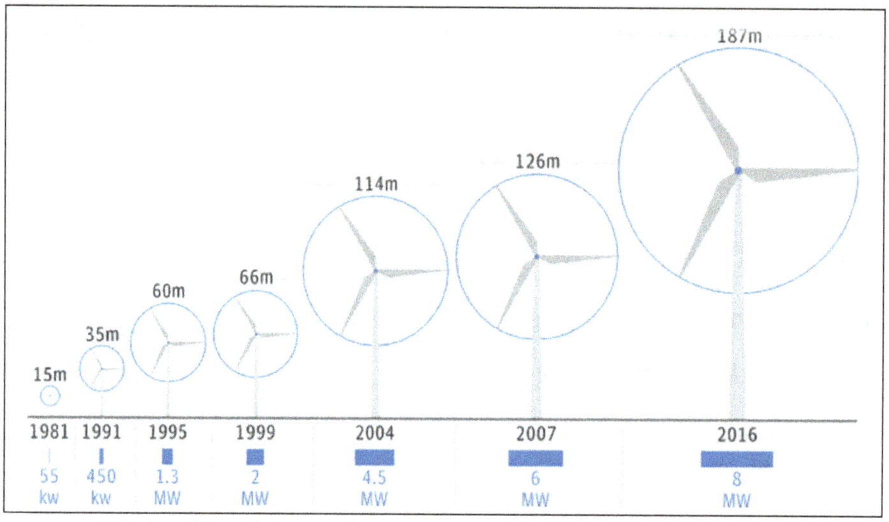

Wind turbine design and power output progression over the years.

Effect of Wind Turbine on Grid

The variable wind speed fluctuate the output voltage and power. It leads to:

- Flicker effects,
- Voltage asymmetry.

This affects the power quality of the network or grid system. Fixed-speed turbines produce a power pulsation emanating from the wind share over height. So variable-wind speed turbines are helpful as they are able to absorb short-term power fluctuations by using immediate storage of energy on the rotating drive train.

Controlling Parameters for Power Generation of Wind Turbine

The process of controlling the velocity of the wind turbine is known as pitching. The speed of wind turbine is controlled by:

- Changing the orientation of wind turbine.
- Pitch of the blades.

This helps to modify the aerodynamics and efficiency of wind turbine.

Components of Wind Turbine

Wind Turbines can be classified in two categories based on rotor structure. Vertical axis wind turbines have a main shaft that stands perpendicular to the direction of the wind stream. Horizontal axis wind turbines have a main shaft that lies along the direction of the wind stream. The experiments and theories contained in this handbook generally pertain to horizontal axis wind turbines. Several wind turbine emulators are used in these experiments, including full-system, lab-scale wind turbines, electric generators, and various measurement systems.

A wind turbine is a system of systems. Each has a particular function, and can be generally classified according to table.

Table: Subsystems of a typical wind turbine generator.

System	Function
Yaw	Track incoming wind direction.
Pitch	Control blade position.
Drivetrain	Shift torque and speed characteristic.
Generator	Convert from mechanical to electrical energy.
Power system interconnection	Interface generator with load or power grid.
SCADA	Monitor performance, control set-points, human interface.

Each system has dependency on the others. Therefore, it is necessary for the turbine to have a system-wide controller to communicate with and coordinate control of various turbine components. Based on information from various sensors, the main controller can set operating conditions, verify performance metrics, and communicate with external parties, including a park-wide supervisory control and data acquisition (SCADA) system.

Yaw

Components around the yaw system of a MW-class turbine.

To keep the wind turbine pointed into the wind, signals from a wind vane (or other wind direction measuring device) are monitored to check incoming wind direction. With this information, the controller can actuate yaw motors to turn the nacelle as necessary. However, many turbine designs are restricted in their yaw movement. Cables that carry power and/or control signals from down-tower to up-tower are generally bundled together, and allowed to twist a specified amount as the nacelle rotates. If those cables are twisted too much, they can be pulled off of their anchors resulting in extreme damage. A limit switch is useful to notify the controller when the twist limit has been reached. Even with this limitation in place, cables can still wear out due to repetitive movement. Yaw system components of a utility-scale wind turbine are pictured in figure.

Another yaw system composed of one motor and one gear-set is shown in figure.

Yaw system drive of a kW-class turbine.

Pitch

Wind turbine blades provide lift and drag forces, similar to an airplane. As air passes around the blades a torque is applied to the main shaft making it accelerate. If no energy were extracted from the system via the generator, and the entire system were lossless, the turbine shaft would accelerate indefinitely. In a real system, turbulence is created around the blades as they cut through the flowing air mass. As the rotor speed increases, the blades will begin to cut into the turbulent air created by the previous blade, causing it to "stall".

Interestingly, stall provides a means of mechanical speed limitation; blades can be designed to stall at specific speeds and rates. More complex turbine designs include individual pitch control, via electric motors or hydraulic cylinders. These systems rely on feedback of blade angle measurements; this can be from proximity sensors on the blade-connecting bolts or encoders on actuating devices. Some of the internal hub components of a hydraulic pitch system are pictured in figure.

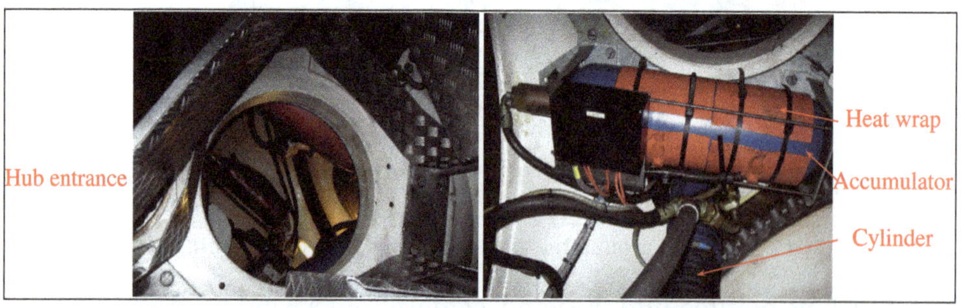

Pitch system components; a hydraulic system in a harsh environment.

Pitch angle directly impacts performance coefficient, a measure of turbine efficiency. Variable pitch angle allows maximization of energy capture and limitation at high-wind. Maximum energy extraction is obtained by operating the blades at a specific aerodynamic condition and the generator at a specific torque and speed condition; both are impacted by the tip-speed ratio. The tip-speed ratio is the steady-state speed at which the blades rotate for a given wind speed. It is defined as:

$$\lambda = \frac{\upsilon_{\text{tip}}}{\upsilon_{\text{wind}}}$$

The tip-speed ratio and corresponding performance coefficient sets the torque vs. speed curve for the wind turbine. By defining the desired tip-speed ratio, a specific toque vs. speed curve can be created; notably, one that maximizes performance. Experiments will be performed later in this handbook to measure performance coefficient across the desired wind speed range to find a desired tip-speed ratio.

The performance coefficient, C_p, is a measure of how much energy is extracted from the turbine compared to how much is available. This measure has a theoretical limit, the Betz limit; $C_{p,\max}$ = 13/27; derivation is provided in your class notes. Do not confuse this with the "capacity factor", which is a merit of the wind resource. Performance coefficient varies with pitch angle and operating speed; this will be observed in a later experiment. It is defined as:

$$C_p(\lambda,\theta) = \frac{P_{gen}(\lambda,\theta)}{0.5\rho A\upsilon^3_{wind}}$$

Where, P_{gen} is the generator terminal power after consideration of all losses and which varies with pitch angle, θ, and tip-speed ratio. A is the swept area and υ_{wind} is the velocity of the incoming wind stream. A typical relationship of performance coefficient over a range of pitch angle and tip-speed ratio is plotted in figure.

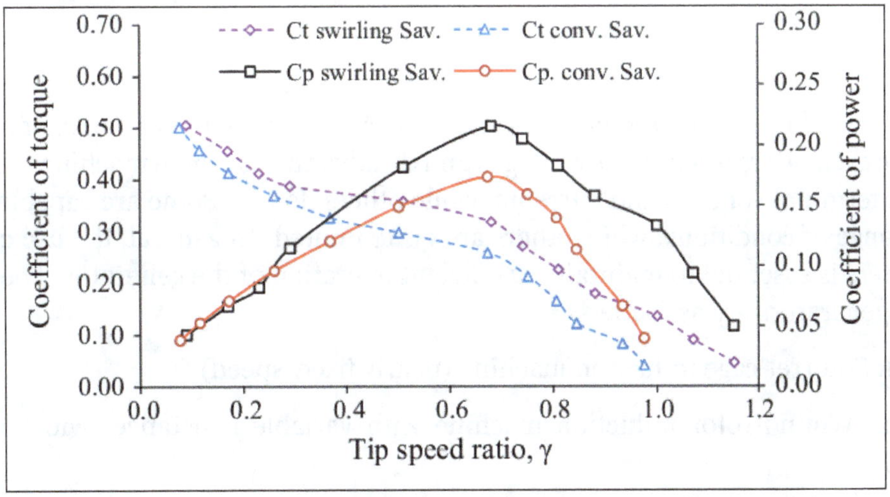

Variation of performance coefficient over tip-speed ratio and rotor speed.

To operate with maximum power extraction, the turbine should operate with the tip-speed ratio and blade angle that maximizes extraction. When operating at high-wind conditions so the blades are pitched back, the tip-speed ratio is adjusted to maintain maximum extraction with the new blade angle. From the plot of figure it's easy to see that the performance coefficient for the turbine that resulted in that data has its performance maximized when $\lambda = 8$ with the blades pitched to $1°$.

Drivetrain

The drivetrain consists of all components attached to energy-transmitting shafts. This includes the main bearing and low-speed shaft, all gearbox shafts and bearings, and the generator shaft assembly which includes a flexible coupling to allow slight shaft misalignment.

The gearbox increases the speed of the shaft connected to the generator. The generators torque and speed characteristics will influence the choice of gear ratio, so that the desired operating wind speed range aligns with a desired generator operating speed range. For doubly-fed induction generators (DFIGs) the gearbox ratio is generally chosen so the DFIG experiences a specific slip-range. For squirrel-cage induction generators, the gear ratio is chosen so that wind speed range is beyond the synchronous speed, putting the squirrel-cage machine in the power generating slip region. For permanent magnet generators, the gear ratio can be chosen to increase the speed of the shaft, which has a direct influence on the operating voltage and efficiency of the generator. Alternatively, wind turbines with permanent magnet generators can be operated in a direct-drive configuration, in which the gearbox is omitted and the generator shaft is directly coupled to the hub, reducing the size and weight of the nacelle, increasing overall efficiency, and reduce the number of moving parts and potential for failure.

Shaft couplings usually include at least one joint which offers flexibility. Without a flexible shaft coupling, vibrations caused by misalignment are transferred through the rigid coupling and into the connected equipment. The result is increased stress on bearing races and rollers. Increased harmonic content is also noticed when operating without a flexible coupling. Imagine gears meshing, causing small impulses, which in turn lead to an impulse response. Unstable nodes can more easily be excited when damping (flex of the coupling) is removed. For these reasons, it is important to take care in ensuring shaft alignment, component integrity, and ample damping.

Generator

Wind turbines are classified by the type of generator. Although there is a range of machines that can do the job, each offers different advantages and disadvantages. Some machines require precise control of the terminal voltages and currents, while others do not. Some are capable of operating over a wide range of conditions, while others are quite limited. In general, the use of power electronic converters is essential to maintaining efficient operation of the generator. The four types of wind turbine generators are as follows:

- Type 1: Squirrel-cage induction machine (nearly fixed-speed).
- Type 2: Wound-rotor induction machine with variable resistance (wider speed/torque range).
- Type 3: Wound-rotor induction machine with power electronics (Doubly-fed, widest range).
- Type 4: Full-converter interfaced machine (typically PMSG or other synchronous constructions).

An example of each type is pictured in figure. Note use of the term "machine"; each of these machines can be operated as either a motor or generator. The difference is whether power is applied to the shaft, or extracted from it; the gear-ratio is such that the desired torque/speed characteristic is reached, or in the case of controlled interfaces, the power converter is programmed to follow a desired torque/speed curve.

The generator supplied with the wind turbine simulator (WTS) illustrating this handbook is an external-field synchronous machine (basically a car alternator). It is a type 4 configuration with the power converter interface being made of a simple rectifier circuit; a boost converter circuit can be added to the output to control generator torque. This is a common machine, and is very popular among amateur wind turbine enthusiasts for its durability and ease of use. However, it is interesting to note that performance of the turbine is greatly affected by the way the generator is operated. The WTS generator is configured to run in an uncontrolled manner, with the stator terminals rectified to provide a dc voltage and current; this generator has a voltage range of 6 – 25 V, depending on the speed and field strength. Several experiments are included later in this handbook to illustrate performance dependence on control technique.

Power System Interconnection

Wind turbines can be operated as part of an existing power distribution network or in a stand-alone island power system. Both require use of controllers, transformers, filters, relays, and other sensors and protective devices. A portion of a DFIG power system interface, including overcurrent and synchronization hardware, is pictured in figure.

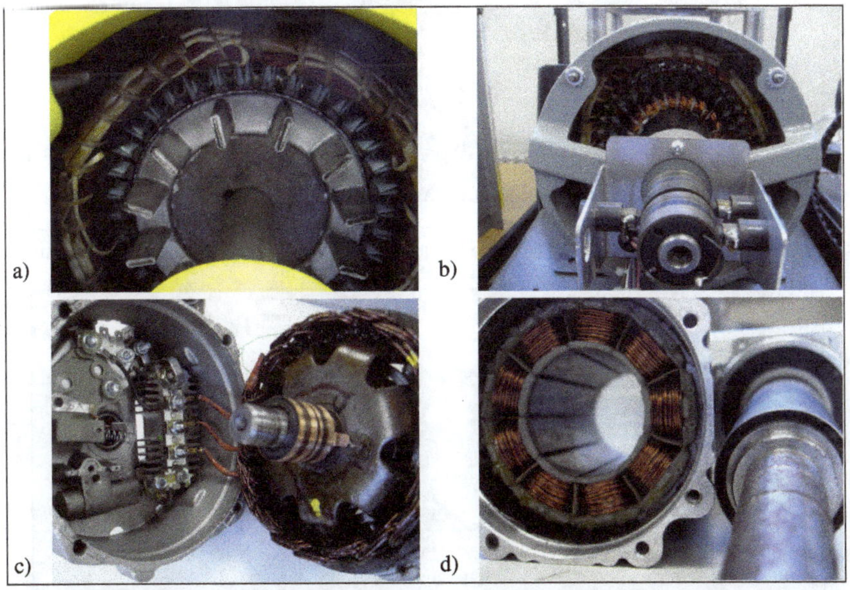

Types of machines used in wind turbines. (a) Type 1 squirrel-cage machine. (b) Type 2 and 3 wound-rotor induction machine. (c) Type 4 external-field synchronous machine. (d) Type 4 permanent magnet machine.

When used in an island mode, the wind turbine plays a large role in stability of the power system. Frequency and voltage can be regulated by the wind turbine or auxiliary systems. Such ancillary services provide reactive power for voltage support, and inertia emulation for frequency regulation. Experiments are performed later in this handbook using a doubly-fed induction generator to illustrate power system interconnection.

The wind turbine must also be capable of limiting its power output, and must have the capability of withstanding various fault conditions. For instance, when connecting a wind turbine to an existing power distribution network, it is required that the terminal voltage and frequency match that of the power system, and that harmonic currents are limited. Filters are useful for harmonic mitigation, and controllers provide automated synchronization.

Supervisory Control and Data Acquisition

Supervisory control and data acquisition (SCADA) systems are an important item to consider. SCADA systems collect information from wind turbines, substations, loads, and system operators, and can control turbine set-points to maintain reliable operation. When power generation signals are provided by a system operator, such as Mid-Continental Independent Service Operator (MISO), the SCADA system receives those and adjusts set-points of individual turbines. It can also shut down turbines in case of excess energy production and emergency operations.

SCADA systems also provide the operator with visual information regarding turbine status and component health. Interfaces are usually provided to visualize system details and provide remote control the wind turbine. An example SCADA screen is shown in figure.

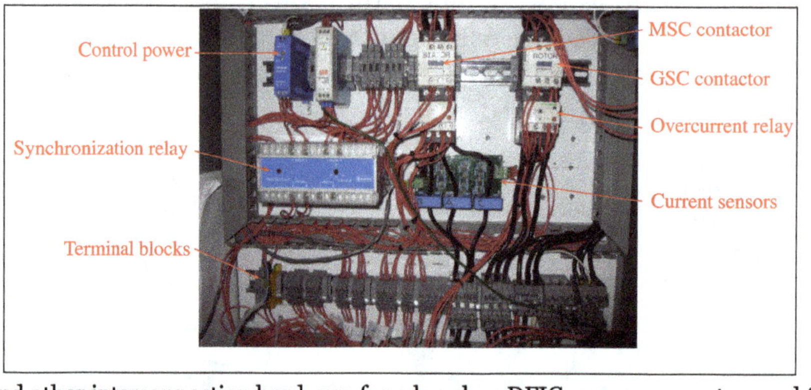

Circuit breakers and other interconnection hardware for a kw-class DFIG; power converters and filters not shown.

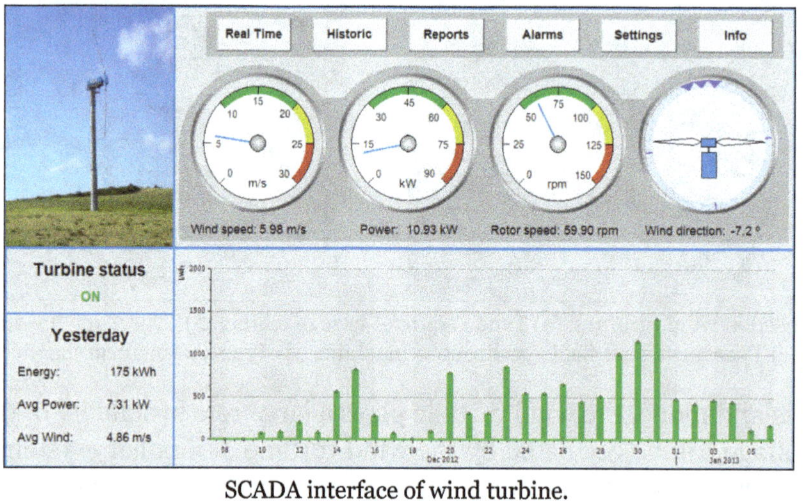

SCADA interface of wind turbine.

Wind Turbine Rotation

1. Wind (moving air that contains kinetic energy) blows toward the turbine's rotor blades.

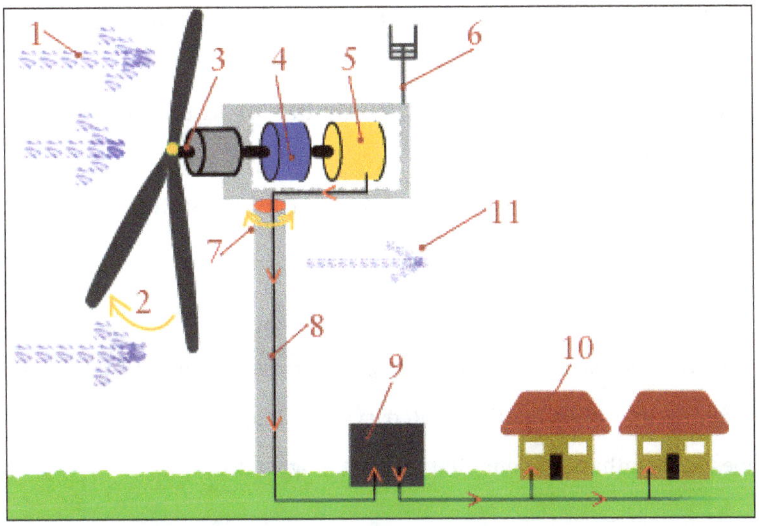

2. The rotors spin around, capturing some of the kinetic energy from the wind, and turning the central drive shaft that supports them. Although the outer edges of the rotor blades move very fast, the central axle (drive shaft) they're connected to turns quite slowly.

3. In most large modern turbines, the rotor blades can swivel on the hub at the front so they meet the wind at the best angle (or "pitch") for harvesting energy. This is called the pitch control mechanism. On big turbines, small electric motors or hydraulic rams swivel the blades back and forth under precise electronic control. On smaller turbines, the pitch control is often completely mechanical. However, many turbines have fixed rotors and no pitch control at all.

4. Inside the nacelle (the main body of the turbine sitting on top of the tower and behind the blades), the gearbox converts the low-speed rotation of the drive shaft (perhaps, 16 revolutions per minute, rpm) into high-speed (perhaps, 1600 rpm) rotation fast enough to drive the generator efficiently.

5. The generator, immediately behind the gearbox, takes kinetic energy from the spinning drive shaft and turns it into electrical energy. Running at maximum capacity, a typical 2MW turbine generator will produce 2 million watts of power at about 700 volts.

6. Anemometers (automatic speed measuring devices) and wind vanes on the back of the nacelle provide measurements of the wind speed and direction.

7. Using these measurements, the entire top part of the turbine (the rotors and nacelle) can be rotated by a yaw motor, mounted between the nacelle and the tower, so it faces directly into the oncoming wind and captures the maximum amount of energy. If it's too windy or turbulent, brakes are applied to stop the rotors from turning (for safety reasons). The brakes are also applied during routine maintenance.

8. The electric current produced by the generator flows through a cable running down through the inside of the turbine tower.

9. A step-up transformer converts the electricity to about 50 times higher voltage so it can be transmitted efficiently to the power grid (or to nearby buildings or communities). If the

electricity is flowing to the grid, it's converted to an even higher voltage (130,000 volts or more) by a substation nearby, which services many turbines.

10. Homes enjoy clean, green energy: the turbine has produced no greenhouse gas emissions or pollution as it operates.
11. Wind carries on blowing past the turbine, but with less speed and energy and more turbulence (since the turbine has disrupted its flow).

References

- Pros-and-cons-of-wind-energy: conserve-energy-future.com, Retrieved 10, February 2020
- What-is-a-wind-farm: wonderopolis.org, Retrieved 27, May 2020
- Windturbines: explainthatstuff.com, Retrieved 07, August 2020

Sun as a Source of Energy

3

Solar energy is the energy that is generated from the Sun in the form of radiant light and heat. The process of conversion of this energy from the Sun is known as solar power. Solar thermal energy is the collection of the Sun's heat for human use. This chapter discusses all the topics related to solar energy like solar thermal power plants, active solar, solar power, etc. in a comprehensive manner.

Solar Energy

Solar energy is radiant light and heat from the Sun that is harnessed using a range of ever-evolving technologies such as solar heating, photovoltaics, solar thermal energy, solar architecture, molten salt power plants and artificial photosynthesis.

The source of Earth's solar power: the Sun.

It is an important source of renewable energy and its technologies are broadly characterized as either passive solar or active solar depending on how they capture and distribute solar energy or convert it into solar power. Active solar techniques include the use of photovoltaic systems, concentrated solar power and solar water heating to harness the energy. Passive solar techniques include orienting a building to the Sun, selecting materials with favorable thermal mass or light-dispersing properties, and designing spaces that naturally circulate air.

The large magnitude of solar energy available makes it a highly appealing source of electricity. The United Nations Development Programme in its 2000 World Energy Assessment found that the annual potential of solar energy was 1,575–49,837 exajoules (EJ). This is several times larger than the total world energy consumption, which was 559.8 EJ in 2012.

In 2011, the International Energy Agency said that "the development of affordable, inexhaustible and clean solar energy technologies will have huge longer-term benefits. It will increase countries' energy security through reliance on an indigenous, inexhaustible and mostly import-independent resource, enhance sustainability, reduce pollution, lower the costs of mitigating global warming, and keep fossil fuel prices lower than otherwise. These advantages are global. Hence the additional costs of the incentives for early deployment should be considered learning investments; they must be wisely spent and need to be widely shared".

Potential

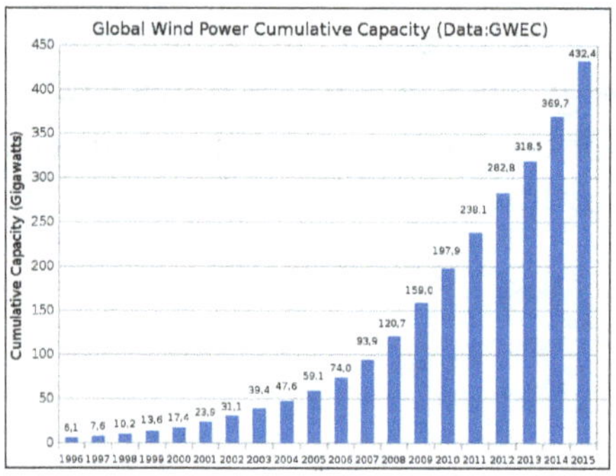

About half the incoming solar energy reaches the Earth's surface.

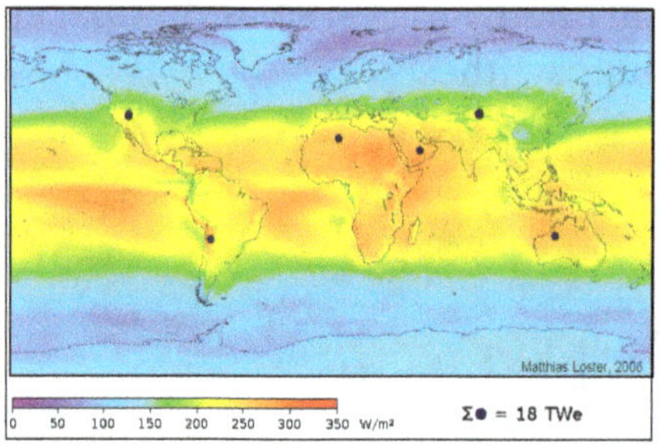

Average insolation. The theoretical area of the small black dots is sufficient to supply the world's total energy needs of 18 TW with solar power.

The Earth receives 174 petawatts (TW) of incoming solar radiation (insolation) at the upper atmosphere. Approximately 30% is reflected back to space while the rest is absorbed by clouds, oceans and land masses. The spectrum of solar light at the Earth's surface is mostly spread across the visible and near-infrared ranges with a small part in the near-ultraviolet. Most of the world's population live in areas with insolation levels of 150-300 watts/m^2, or 3.5-7.0 kWh/m^2 per day.

Solar radiation is absorbed by the Earth's land surface, oceans – which cover about 71% of the globe – and atmosphere. Warm air containing evaporated water from the oceans rises, causing atmospheric circulation or convection. When the air reaches a high altitude, where the temperature is low, water vapor condenses into clouds, which rain onto the Earth's surface, completing the water cycle. The latent heat of water condensation amplifies convection, producing atmospheric phenomena such as wind, cyclones and anti-cyclones. Sunlight absorbed by the oceans and land masses keeps the surface at an average temperature of 14 °C. By photosynthesis, green plants convert solar energy into chemically stored energy, which produces food, wood and the biomass from which fossil fuels are derived.

The total solar energy absorbed by Earth's atmosphere, oceans and land masses is approximately 3,850,000 exajoules (EJ) per year. In 2002, this was more energy in one hour than the world used in one year. Photosynthesis captures approximately 3,000 EJ per year in biomass. The amount of solar energy reaching the surface of the planet is so vast that in one year it is about twice as much as will ever be obtained from all of the Earth's non-renewable resources of coal, oil, natural gas, and mined uranium combined,

Yearly solar fluxes & human consumption[1]		
Solar	3,850,000	
Wind	2,250	
Biomass potential	~200	
Primary energy use[2]	539	
Electricity[2]	~67	
[1] Energy given in Exajoule (EJ) = 10^{18} J = 278 TWh [2] Consumption as of year 2010		

The potential solar energy that could be used by humans differs from the amount of solar energy present near the surface of the planet because factors such as geography, time variation, cloud cover, and the land available to humans limit the amount of solar energy that we can acquire.

Geography affects solar energy potential because areas that are closer to the equator have a greater amount of solar radiation. However, the use of photovoltaics that can follow the position of the sun can significantly increase the solar energy potential in areas that are farther from the equator. Time variation effects the potential of solar energy because during the nighttime there is little solar radiation on the surface of the Earth for solar panels to absorb. This limits the amount of energy that solar panels can absorb in one day. Cloud cover can affect the potential of solar panels because clouds block incoming light from the sun and reduce the light available for solar cells.

In addition, land availability has a large effect on the available solar energy because solar panels can only be set up on land that is otherwise unused and suitable for solar panels. Roofs have been found to be a suitable place for solar cells, as many people have discovered that they can collect energy directly from their homes this way. Other areas that are suitable for solar cells are lands that are not being used for businesses where solar plants can be established.

Solar technologies are characterized as either passive or active depending on the way they capture, convert and distribute sunlight and enable solar energy to be harnessed at different levels around the world, mostly depending on distance from the equator. Although solar energy refers primarily to the use of solar radiation for practical ends, all renewable energies, other than Geothermal power and Tidal power, derive their energy either directly or indirectly from the Sun.

Active solar techniques use photovoltaics, concentrated solar power, solar thermal collectors, pumps, and fans to convert sunlight into useful outputs. Passive solar techniques include selecting materials with favorable thermal properties, designing spaces that naturally circulate air, and referencing the position of a building to the Sun. Active solar technologies increase the supply of energy and are considered supply side technologies, while passive solar technologies reduce the need for alternate resources and are generally considered demand side technologies.

In 2000, the United Nations Development Programme, UN Department of Economic and Social Affairs, and World Energy Council published an estimate of the potential solar energy that could be used by humans each year that took into account factors such as insolation, cloud cover, and the land that is usable by humans. The estimate found that solar energy has a global potential of 1,575–49,837 EJ per year.

Annual solar energy potential by region (Exajoules)												
Region	North America	Latin America and Caribbean	Western Europe	Central and Eastern Europe	Former Soviet Union	Middle East and North Africa	Sub-Saharan Africa	Pacific Asia	South Asia	Centrally planned Asia	Pacific OECD
Minimum	181.1	112.6	25.1	4.5	199.3	412.4	371.9	41.0	38.8	115.5	72.6
Maximum	7,410	3,385	914	154	8,655	11,060	9,528	994	1,339	4,135	2,263

Quantitative relation of global solar potential vs. the world's primary energy consumption:

- Ratio of potential vs. current consumption (402 EJ) as of year: 3.9 (minimum) to 124 (maximum).
- Ratio of potential vs. projected consumption by 2050 (590–1,050 EJ): 1.5–2.7 (minimum) to 47–84 (maximum).
- Ratio of potential vs. projected consumption by 2100 (880–1,900 EJ): 0.8–1.8 (minimum) to 26–57 (maximum).

Thermal Energy

Solar thermal technologies can be used for water heating, space heating, space cooling and process heat generation.

Early Commercial Adaptation

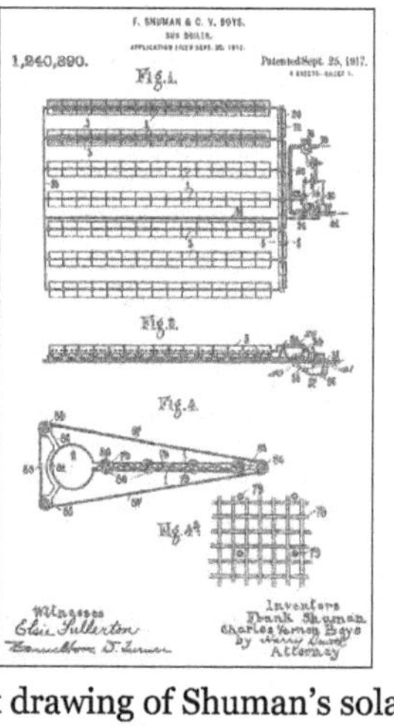

1917 Patent drawing of Shuman's solar collector.

In 1897, Frank Shuman, a U.S. inventor, engineer and solar energy pioneer built a small demonstration solar engine that worked by reflecting solar energy onto square boxes filled with ether, which has a lower boiling point than water, and were fitted internally with black pipes which in turn powered a steam engine. In 1908 Shuman formed the Sun Power Company with the intent of building larger solar power plants. He, along with his technical advisor A.S.E. Ackermann and

British physicist Sir Charles Vernon Boys, developed an improved system using mirrors to reflect solar energy upon collector boxes, increasing heating capacity to the extent that water could now be used instead of ether. Shuman then constructed a full-scale steam engine powered by low-pressure water, enabling him to patent the entire solar engine system by 1912.

Shuman built the world's first solar thermal power station in Maadi, Egypt, between 1912 and 1913. His plant used parabolic troughs to power a 45–52 kilowatts (60–70 hp) engine that pumped more than 22,000 litres (4,800 imp gal; 5,800 US gal) of water per minute from the Nile River to adjacent cotton fields. Although the outbreak of World War I and the discovery of cheap oil in the 1930s discouraged the advancement of solar energy, Shuman's vision and basic design were resurrected in the 1970s with a new wave of interest in solar thermal energy. In 1916 Shuman was quoted in the media advocating solar energy's utilization, saying.

We have proved the commercial profit of sun power in the tropics and have more particularly proved that after our stores of oil and coal are exhausted the human race can receive unlimited power from the rays of the sun.

—Frank Shuman, New York Times, 2 July 1916

Water Heating

Solar water heaters facing the Sun to maximize gain.

Solar hot water systems use sunlight to heat water. In low geographical latitudes (below 40 degrees) from 60 to 70% of the domestic hot water use with temperatures up to 60 °C can be provided by solar heating systems. The most common types of solar water heaters are evacuated tube collectors (44%) and glazed flat plate collectors (34%) generally used for domestic hot water; and unglazed plastic collectors (21%) used mainly to heat swimming pools.

As of 2007, the total installed capacity of solar hot water systems was approximately 154 thermal gigawatt (GW_{th}). China is the world leader in their deployment with 70 GW_{th} installed as of 2006 and a long-term goal of 210 GW_{th} by 2020. Israel and Cyprus are the per capita leaders in the use of solar hot water systems with over 90% of homes using them. In the United States, Canada, and Australia, heating swimming pools is the dominant application of solar hot water with an installed capacity of 18 GW_{th} as of 2005.

Heating, Cooling and Ventilation

In the United States, heating, ventilation and air conditioning (HVAC) systems account for 30% (4.65 EJ/yr) of the energy used in commercial buildings and nearly 50% (10.1 EJ/yr) of the energy used in residential buildings. Solar heating, cooling and ventilation technologies can be used to offset a portion of this energy.

MIT's Solar House #1, built in 1939 in the U.S., used seasonal thermal energy storage for year-round heating.

Thermal mass is any material that can be used to store heat—heat from the Sun in the case of solar energy. Common thermal mass materials include stone, cement and water. Historically they have been used in arid climates or warm temperate regions to keep buildings cool by absorbing solar energy during the day and radiating stored heat to the cooler atmosphere at night. However, they can be used in cold temperate areas to maintain warmth as well. The size and placement of thermal mass depend on several factors such as climate, daylighting and shading conditions. When properly incorporated, thermal mass maintains space temperatures in a comfortable range and reduces the need for auxiliary heating and cooling equipment.

A solar chimney (or thermal chimney, in this context) is a passive solar ventilation system composed of a vertical shaft connecting the interior and exterior of a building. As the chimney warms, the air inside is heated causing an updraft that pulls air through the building. Performance can be improved by using glazing and thermal mass materials in a way that mimics greenhouses.

Deciduous trees and plants have been promoted as a means of controlling solar heating and cooling. When planted on the southern side of a building in the northern hemisphere or the northern side in the southern hemisphere, their leaves provide shade during the summer, while the bare limbs allow light to pass during the winter. Since bare, leafless trees shade 1/3 to 1/2 of incident solar radiation, there is a balance between the benefits of summer shading and the corresponding loss of winter heating. In climates with significant heating loads, deciduous trees should not be planted on the Equator-facing side of a building because they will interfere with winter solar availability. They can, however, be used on the east and west sides to provide a degree of summer shading without appreciably affecting winter solar gain.

Cooking

Parabolic dish produces steam for cooking, in Auroville, India.

Solar cookers use sunlight for cooking, drying and pasteurization. They can be grouped into three broad categories: box cookers, panel cookers and reflector cookers. The simplest solar cooker is the box cooker first built by Horace de Saussure in 1767. A basic box cooker consists of an insulated container with a transparent lid. It can be used effectively with partially overcast skies and will typically reach temperatures of 90–150 °C (194–302 °F). Panel cookers use a reflective panel to direct sunlight onto an insulated container and reach temperatures comparable to box cookers. Reflector cookers use various concentrating geometries (dish, trough, Fresnel mirrors) to focus light on a cooking container. These cookers reach temperatures of 315 °C (599 °F) and above but require direct light to function properly and must be repositioned to track the Sun.

Process Heat

Solar concentrating technologies such as parabolic dish, trough and Scheffler reflectors can provide process heat for commercial and industrial applications. The first commercial system was the Solar Total Energy Project (STEP) in Shenandoah, Georgia, USA where a field of 114 parabolic dishes provided 50% of the process heating, air conditioning and electrical requirements for a clothing factory. This grid-connected cogeneration system provided 400 kW of electricity plus thermal energy in the form of 401 kW steam and 468 kW chilled water, and had a one-hour peak load thermal storage. Evaporation ponds are shallow pools that concentrate dissolved solids through evaporation. The use of evaporation ponds to obtain salt from seawater is one of the oldest applications of solar energy. Modern uses include concentrating brine solutions used in leach mining and removing dissolved solids from waste streams. Clothes lines, clotheshorses, and clothes racks dry clothes through evaporation by wind and sunlight without consuming electricity or gas. In some states of the United States legislation protects the "right to dry" clothes. Unglazed transpired collectors (UTC) are perforated sun-facing walls used for preheating ventilation air. UTCs can raise the incoming air temperature up to 22 °C (40 °F) and deliver outlet temperatures of 45–60 °C (113–140 °F). The short payback period of transpired collectors (3 to 12 years) makes them a more cost-effective alternative than glazed collection systems. As of 2003, over 80 systems with a combined collector area of 35,000 square metres (380,000 sq ft) had been installed worldwide, including an 860 m^2 (9,300 sq ft) collector in Costa Rica used for drying coffee beans and a 1,300 m^2 (14,000 sq ft) collector in Coimbatore, India, used for drying marigolds.

Water Treatment

Solar water disinfection in Indonesia.

Solar distillation can be used to make saline or brackish water potable. The first recorded instance of this was by 16th-century Arab alchemists. A large-scale solar distillation project was first constructed in 1872 in the Chilean mining town of Las Salinas. The plant, which had solar collection area of 4,700 m² (51,000 sq ft), could produce up to 22,700 L (5,000 imp gal; 6,000 US gal) per day and operate for 40 years. Individual still designs include single-slope, double-slope (or greenhouse type), vertical, conical, inverted absorber, multi-wick, and multiple effect. These stills can operate in passive, active, or hybrid modes. Double-slope stills are the most economical for decentralized domestic purposes, while active multiple effect units are more suitable for large-scale applications.

Solar water disinfection (SODIS) involves exposing water-filled plastic polyethylene terephthalate (PET) bottles to sunlight for several hours. Exposure times vary depending on weather and climate from a minimum of six hours to two days during fully overcast conditions. It is recommended by the World Health Organization as a viable method for household water treatment and safe storage. Over two million people in developing countries use this method for their daily drinking water.

Solar energy may be used in a water stabilization pond to treat waste water without chemicals or electricity. A further environmental advantage is that algae grow in such ponds and consume carbon dioxide in photosynthesis, although algae may produce toxic chemicals that make the water unusable.

Molten Salt Technology

Molten salt can be employed as a thermal energy storage method to retain thermal energy collected by a solar tower or solar trough of a concentrated solar power plant, so that it can be used to generate electricity in bad weather or at night. It was demonstrated in the Solar Two project from 1995–1999. The system is predicted to have an annual efficiency of 99%, a reference to the energy retained by storing heat before turning it into electricity, versus converting heat directly into electricity. The molten salt mixtures vary. The most extended mixture contains sodium nitrate, potassium nitrate and calcium nitrate. It is non-flammable and nontoxic, and has already been used in the chemical and metals industries as a heat-transport fluid, so experience with such systems exists in non-solar applications.

The salt melts at 131 °C (268 °F). It is kept liquid at 288 °C (550 °F) in an insulated "cold" storage tank. The liquid salt is pumped through panels in a solar collector where the focused sun heats it to 566 °C (1,051 °F). It is then sent to a hot storage tank. This is so well insulated that the thermal energy can be usefully stored for up to a week.

When electricity is needed, the hot salt is pumped to a conventional steam-generator to produce superheated steam for a turbine/generator as used in any conventional coal, oil, or nuclear power plant. A 100-megawatt turbine would need a tank about 9.1 metres (30 ft) tall and 24 metres (79 ft) in diameter to drive it for four hours by this design.

Several parabolic trough power plants in Spain and solar power tower developer SolarReserve use this thermal energy storage concept. The Solana Generating Station in the U.S. has six hours of storage by molten salt. The María Elena plant is a 400 MW thermo-solar complex in the northern Chilean region of Antofagasta employing molten salt technology.

Electricity Production

Some of the world's largest solar power stations: Ivanpah (CSP) and Topaz (PV).

Solar power is the conversion of sunlight into electricity, either directly using photovoltaics (PV), or indirectly using concentrated solar power (CSP). CSP systems use lenses or mirrors and tracking systems to focus a large area of sunlight into a small beam. PV converts light into electric current using the photoelectric effect.

Solar power is anticipated to become the world's largest source of electricity by 2050, with solar photovoltaics and concentrated solar power contributing 16 and 11 percent to the global overall consumption, respectively.

Commercial CSP plants were first developed in the 1980s. Since 1985 the eventually 354 MW SEGS CSP installation, in the Mojave Desert of California, is the largest solar power plant in the world. Other large CSP plants include the 150 MW Solnova Solar Power Station and the 100 MW Andasol solar power station, both in Spain. The 250 MW Agua Caliente Solar Project, in the United

States, and the 221 MW Charanka Solar Park in India, are the world's largest photovoltaic plants. Solar projects exceeding 1 GW are being developed, but most of the deployed photovoltaics are in small rooftop arrays of less than 5 kW, which are connected to the grid using net metering and/or a feed-in tariff. In 2013 solar generated less than 1% of the world's total grid electricity.

Photovoltaics

In the last two decades, photovoltaics (PV), also known as solar PV, has evolved from a pure niche market of small scale applications towards becoming a mainstream electricity source. A solar cell is a device that converts light directly into electricity using the photoelectric effect. The first solar cell was constructed by Charles Fritts in the 1880s. In 1931 a German engineer, Dr Bruno Lange, developed a photo cell using silver selenide in place of copper oxide. Although the prototype selenium cells converted less than 1% of incident light into electricity, both Ernst Werner von Siemens and James Clerk Maxwell recognized the importance of this discovery. Following the work of Russell Ohl in the 1940s, researchers Gerald Pearson, Calvin Fuller and Daryl Chapin created the crystalline silicon solar cell in 1954. These early solar cells cost 286 USD/watt and reached efficiencies of 4.5–6%. By 2012 available efficiencies exceeded 20%, and the maximum efficiency of research photovoltaics was in excess of 40%.

Concentrated Solar Power

Concentrating Solar Power (CSP) systems use lenses or mirrors and tracking systems to focus a large area of sunlight into a small beam. The concentrated heat is then used as a heat source for a conventional power plant. A wide range of concentrating technologies exists; the most developed are the parabolic trough, the concentrating linear fresnel reflector, the Stirling dish and the solar power tower. Various techniques are used to track the Sun and focus light. In all of these systems a working fluid is heated by the concentrated sunlight, and is then used for power generation or energy storage.

Architecture and Urban Planning

Darmstadt University of Technology, Germany, won the 2007 Solar Decathlon in Washington, D.C. with this passive house designed for humid and hot subtropical climate.

Sunlight has influenced building design since the beginning of architectural history. Advanced solar architecture and urban planning methods were first employed by the Greeks and Chinese, who oriented their buildings toward the south to provide light and warmth.

The common features of passive solar architecture are orientation relative to the Sun, compact proportion (a low surface area to volume ratio), selective shading (overhangs) and thermal mass. When these features are tailored to the local climate and environment they can produce well-lit spaces that stay in a comfortable temperature range. Socrates' Megaron House is a classic example of passive solar design. The most recent approaches to solar design use computer modeling tying together solar lighting, heating and ventilation systems in an integrated solar design package. Active solar equipment such as pumps, fans and switchable windows can complement passive design and improve system performance.

Urban heat islands (UHI) are metropolitan areas with higher temperatures than that of the surrounding environment. The higher temperatures result from increased absorption of solar energy by urban materials such as asphalt and concrete, which have lower albedos and higher heat capacities than those in the natural environment. A straightforward method of counteracting the UHI effect is to paint buildings and roads white, and to plant trees in the area. Using these methods, a hypothetical "cool communities" program in Los Angeles has projected that urban temperatures could be reduced by approximately 3 °C at an estimated cost of US$1 billion, giving estimated total annual benefits of US$530 million from reduced air-conditioning costs and healthcare savings.

Agriculture and Horticulture

Greenhouses like these in the Westland municipality of the Netherlands grow vegetables, fruits and flowers.

Agriculture and horticulture seek to optimize the capture of solar energy in order to optimize the productivity of plants. Techniques such as timed planting cycles, tailored row orientation, staggered heights between rows and the mixing of plant varieties can improve crop yields. While sunlight is generally considered a plentiful resource, the exceptions highlight the importance of solar energy to agriculture. During the short growing seasons of the Little Ice Age, French and English farmers employed fruit walls to maximize the collection of solar energy. These walls acted as thermal masses and accelerated ripening by keeping plants warm. Early fruit walls were built perpendicular to the ground and facing south, but over time, sloping walls were developed to make better use of sunlight. In 1699, Nicolas Fatio de Duillier even suggested using a tracking mechanism which could pivot to follow the Sun. Applications of solar energy in agriculture aside from growing crops include pumping water, drying crops, brooding chicks and drying chicken manure. More recently the technology has been embraced by vintners, who use the energy generated by solar panels to power grape presses.

Greenhouses convert solar light to heat, enabling year-round production and the growth (in enclosed environments) of specialty crops and other plants not naturally suited to the local climate. Primitive greenhouses were first used during Roman times to produce cucumbers year-round for the Roman emperor Tiberius. The first modern greenhouses were built in Europe in the 16th century to keep exotic plants brought back from explorations abroad. Greenhouses remain an important part of horticulture today, and plastic transparent materials have also been used to similar effect in polytunnels and row covers.

Transport

Winner of the 2013 World Solar Challenge in Australia.

Solar electric aircraft circumnavigating the globe in 2015.

Development of a solar-powered car has been an engineering goal since the 1980s. The World Solar Challenge is a biannual solar-powered car race, where teams from universities and enterprises compete over 3,021 kilometres (1,877 mi) across central Australia from Darwin to Adelaide. In 1987, when it was founded, the winner's average speed was 67 kilometres per hour (42 mph) and by 2007 the winner's average speed had improved to 90.87 kilometres per hour (56.46 mph). The North American Solar Challenge and the planned South African Solar Challenge are comparable competitions that reflect an international interest in the engineering and development of solar powered vehicles.

Some vehicles use solar panels for auxiliary power, such as for air conditioning, to keep the interior cool, thus reducing fuel consumption.

In 1975, the first practical solar boat was constructed in England. By 1995, passenger boats incorporating PV panels began appearing and are now used extensively. In 1996, Kenichi Horie made the first solar-powered crossing of the Pacific Ocean, and the *Sun21* catamaran made the first solar-powered crossing of the Atlantic Ocean in the winter of 2006–2007. There were plans to circumnavigate the globe in 2010.

In 1974, the unmanned AstroFlight Sunrise airplane made the first solar flight. On 29 April 1979, the *Solar Riser* made the first flight in a solar-powered, fully controlled, man-carrying flying machine, reaching an altitude of 40 ft (12 m). In 1980, the *Gossamer Penguin* made the first piloted flights powered solely by photovoltaics. This was quickly followed by the *Solar Challenger* which crossed the English Channel in July 1981. In 1990 Eric Scott Raymond in 21 hops flew from California to North Carolina using solar power. Developments then turned back to unmanned aerial vehicles (UAV) with the *Pathfinder* (1997) and subsequent designs, culminating in the *Helios* which set the altitude record for a non-rocket-propelled aircraft at 29,524 metres (96,864 ft) in 2001. The *Zephyr*, developed by BAE Systems, is the latest in a line of record-breaking solar aircraft, making a 54-hour flight in 2007, and month-long flights were envisioned by 2010. As of 2016, Solar Impulse, an electric aircraft, is currently circumnavigating the globe. It is a single-seat plane powered by solar cells and capable of taking off under its own power. The design allows the aircraft to remain airborne for several days.

A solar balloon is a black balloon that is filled with ordinary air. As sunlight shines on the balloon, the air inside is heated and expands causing an upward buoyancy force, much like an artificially heated hot air balloon. Some solar balloons are large enough for human flight, but usage is generally limited to the toy market as the surface-area to payload-weight ratio is relatively high.

Fuel Production

Concentrated solar panels are getting a power boost. Pacific Northwest National Laboratory (PNNL) will be testing a new concentrated solar power system -- one that can help natural gas power plants reduce their fuel usage by up to 20 percent.

Solar chemical processes use solar energy to drive chemical reactions. These processes offset energy that would otherwise come from a fossil fuel source and can also convert solar energy into storable and transportable fuels. Solar induced chemical reactions can be divided into thermochemical or photochemical. A variety of fuels can be produced by artificial photosynthesis. The multielectron catalytic chemistry involved in making carbon-based fuels (such as methanol) from reduction of carbon dioxide is challenging; a feasible alternative is hydrogen production from protons, though use of water as the source of electrons (as plants do) requires mastering the multielectron oxidation of two water molecules to molecular oxygen. Some have envisaged working solar fuel plants in coastal metropolitan areas by 2050 – the splitting of sea water providing hydrogen to be run through adjacent fuel-cell electric power plants and the pure water by-product going directly into the municipal water system. Another vision involves all human structures covering the earth's surface (i.e., roads, vehicles and buildings) doing photosynthesis more efficiently than plants.

Hydrogen production technologies have been a significant area of solar chemical research since the 1970s. Aside from electrolysis driven by photovoltaic or photochemical cells, several thermochemical processes have also been explored. One such route uses concentrators to split water into oxygen and hydrogen at high temperatures (2,300–2,600 °C or 4,200–4,700 °F). Another approach uses the heat from solar concentrators to drive the steam reformation of natural gas thereby increasing the overall hydrogen yield compared to conventional reforming methods. Thermochemical cycles characterized by the decomposition and regeneration of reactants present another avenue for hydrogen production. The Solzinc process under development at the Weizmann Institute of Science uses a 1 MW solar furnace to decompose zinc oxide (ZnO) at temperatures above 1,200 °C (2,200 °F). This initial reaction produces pure zinc, which can subsequently be reacted with water to produce hydrogen.

Energy Storage Methods

Thermal energy storage. The Andasol CSP plant uses tanks of molten salt to store solar energy.

Thermal mass systems can store solar energy in the form of heat at domestically useful temperatures for daily or interseasonal durations. Thermal storage systems generally use readily available materials with high specific heat capacities such as water, earth and stone. Well-designed systems can lower peak demand, shift time-of-use to off-peak hours and reduce overall heating and cooling requirements.

Phase change materials such as paraffin wax and Glauber's salt are another thermal storage medium. These materials are inexpensive, readily available, and can deliver domestically useful temperatures (approximately 64 °C or 147 °F). The "Dover House" (in Dover, Massachusetts) was the first to use a Glauber's salt heating system, in 1948. Solar energy can also be stored at high temperatures using molten salts. Salts are an effective storage medium because they are low-cost, have a high specific heat capacity and can deliver heat at temperatures compatible with conventional power systems. The Solar Two project used this method of energy storage, allowing it to store 1.44 terajoules (400,000 kWh) in its 68 m^3 storage tank with an annual storage efficiency of about 99%.

Off-grid PV systems have traditionally used rechargeable batteries to store excess electricity. With grid-tied systems, excess electricity can be sent to the transmission grid, while standard grid electricity can be used to meet shortfalls. Net metering programs give household systems a credit for any electricity they deliver to the grid. This is handled by 'rolling back' the meter whenever the home produces more electricity than it consumes. If the net electricity use is below zero, the utility then rolls over the kilowatt hour credit to the next month. Other approaches involve the use of two meters, to measure electricity consumed vs. electricity produced. This is less common due to the increased installation cost of the second meter. Most standard meters accurately measure in both directions, making a second meter unnecessary.

Pumped-storage hydroelectricity stores energy in the form of water pumped when energy is available from a lower elevation reservoir to a higher elevation one. The energy is recovered when demand is high by releasing the water, with the pump becoming a hydroelectric power generator.

Development, Deployment and Economics

Participants in a workshop on sustainable development inspect solar panels at Monterrey Institute of Technology and Higher Education, Mexico City on top of a building on campus.

Beginning with the surge in coal use which accompanied the Industrial Revolution, energy consumption has steadily transitioned from wood and biomass to fossil fuels. The early development of solar technologies starting in the 1860s was driven by an expectation that coal would soon become scarce. However, development of solar technologies stagnated in the early 20th century in the face of the increasing availability, economy, and utility of coal and petroleum.

The 1973 oil embargo and 1979 energy crisis caused a reorganization of energy policies around the world and brought renewed attention to developing solar technologies. Deployment strategies focused on incentive programs such as the Federal Photovoltaic Utilization Program in the U.S. and the Sunshine Program in Japan. Other efforts included the formation of research facilities in the U.S. (SERI, now NREL), Japan (NEDO), and Germany (Fraunhofer Institute for Solar Energy Systems ISE).

Commercial solar water heaters began appearing in the United States in the 1890s. These systems saw increasing use until the 1920s but were gradually replaced by cheaper and more reliable heating fuels. As with photovoltaics, solar water heating attracted renewed attention as a result of the oil crises in the 1970s but interest subsided in the 1980s due to falling petroleum prices. Development in the solar water heating sector progressed steadily throughout the 1990s and annual growth rates have averaged 20% since 1999. Although generally underestimated, solar water heating and cooling is by far the most widely deployed solar technology with an estimated capacity of 154 GW as of 2007.

The International Energy Agency has said that solar energy can make considerable contributions to solving some of the most urgent problems the world now faces:

The development of affordable, inexhaustible and clean solar energy technologies will have huge longer-term benefits. It will increase countries' energy security through reliance on an indigenous, inexhaustible and mostly import-independent resource, enhance sustainability, reduce pollution, lower the costs of mitigating climate change, and keep fossil fuel prices lower than otherwise. These advantages are global. Hence the additional costs of the incentives for early deployment should be considered learning investments; they must be wisely spent and need to be widely shared.

In 2011, a report by the International Energy Agency found that solar energy technologies such as photovoltaics, solar hot water and concentrated solar power could provide a third of the world's energy by 2060 if politicians commit to limiting climate change. The energy from the sun could play a key role in de-carbonizing the global economy alongside improvements in energy efficiency and imposing costs on greenhouse gas emitters. "The strength of solar is the incredible variety and flexibility of applications, from small scale to big scale".

We have proved ... that after our stores of oil and coal are exhausted the human race can receive unlimited power from the rays of the sun.

—Frank Shuman, New York Times, 2 July 1916

ISO Standards

The International Organization for Standardization has established several standards relating to solar energy equipment. For example, ISO 9050 relates to glass in building while ISO 10217 relates to the materials used in solar water heaters.

Solar Thermal Energy

Solar thermal energy (STE) is a form of energy and a technology for harnessing solar energy to generate thermal energy or electrical energy for use in industry, and in the residential and commercial sectors.

Roof-mounted close-coupled thermosiphon solar water heater.

The first three units of Solnova in the foreground, with the two towers of the PS10 and PS20 solar power stations in the background.

Overview

Solar thermal collectors are classified by the United States Energy Information Administration as low-, medium-, or high-temperature collectors. Low-temperature collectors are flat plates generally used to heat swimming pools. Medium-temperature collectors are also usually flat plates but are used for heating water or air for residential and commercial use. High-temperature collectors concentrate sunlight using mirrors or lenses and are generally used for fulfilling heat requirements up to 300 deg C / 20 bar pressure in industries, and for electric power production. Two categories include Concentrated Solar Thermal (CST) for fulfilling heat requirements in industries, and Concentrated Solar Power (CSP) when the heat collected is used for power generation. CST and CSP are not replaceable in terms of application. The largest facilities are located in the American Mojave Desert of California and Nevada. These plants employ a variety of different technologies. The largest examples include, Ivanpah Solar Power Facility (377 MW), Solar Energy Generating Systems installation (354 MW), and Crescent Dunes (110 MW). Spain is the other major developer of solar thermal power plant. The largest examples include, Solnova Solar Power Station (150 MW), the Andasol solar power station (150 MW), and Extresol Solar Power Station (100 MW).

History

Augustin Mouchot demonstrated a solar collector with a cooling engine making ice cream at the 1878 Universal Exhibition in Paris. The first installation of solar thermal energy equipment occurred in the Sahara approximately in 1910 by Frank Shuman when a steam engine was run on steam produced by sunlight. Because liquid fuel engines were developed and found more convenient, the Sahara project was abandoned, only to be revisited several decades later.

Low-temperature Solar Heating and Cooling Systems

Systems for utilizing low-temperature solar thermal energy include means for heat collection; usually heat storage, either short-term or interseasonal; and distribution within a structure or a district heating network. In some cases more than one of these functions is inherent to a single feature of the system (e.g. some kinds of solar collectors also store heat). Some systems are passive, others are active (requiring other external energy to function).

Heating is the most obvious application, but solar cooling can be achieved for a building or district cooling network by using a heat-driven absorption or adsorption chiller (heat pump). There is a productive coincidence that the greater the driving heat from insulation, the greater the cooling output. In 1878, Auguste Mouchout pioneered solar cooling by making ice using a solar steam engine attached to a refrigeration device.

MIT's Solar House #1 built in 1939 used seasonal thermal energy storage (STES) for year-round heating.

In the United States, heating, ventilation, and air conditioning (HVAC) systems account for over 25% (4.75 EJ) of the energy used in commercial buildings and nearly half (10.1 EJ) of the energy used in residential buildings. Solar heating, cooling, and ventilation technologies can be used to offset a portion of this energy.

In Europe, since the mid-1990s about 125 large solar-thermal district heating plants have been constructed, each with over 500 m² (5400 ft²) of solar collectors. The largest are about 10,000 m², with capacities of 7 MW-thermal and solar heat costs around 4 Eurocents/kWh without subsidies. 40 of them have nominal capacities of 1 MW-thermal or more. The Solar District Heating program (SDH) has participation from 14 European Nations and the European Commission, and is working toward technical and market development, and holds annual conferences.

Low-temperature Collectors

Glazed solar collectors are designed primarily for space heating. They recirculate building air through a solar air panel where the air is heated and then directed back into the building. These solar space heating systems require at least two penetrations into the building and only perform when the air in the solar collector is warmer than the building room temperature. Most glazed collectors are used in the residential sector.

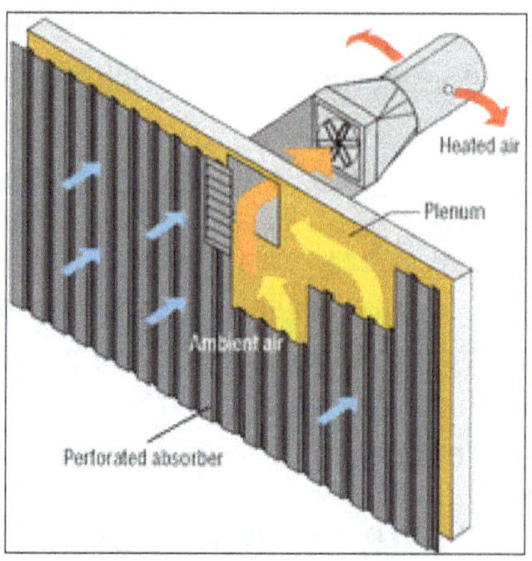

Unglazed, "transpired" air collector.

Unglazed solar collectors are primarily used to pre-heat make-up ventilation air in commercial, industrial and institutional buildings with a high ventilation load. They turn building walls or sections of walls into low cost, high performance, unglazed solar collectors. Also called, "transpired solar panels" or "solar wall", they employ a painted perforated metal solar heat absorber that also serves as the exterior wall surface of the building. Heat conducts from the absorber surface to the thermal boundary layer of air 1 mm thick on the outside of the absorber and to air that passes behind the absorber. The boundary layer of air is drawn into a nearby perforation before the heat can escape by convection to the outside air. The heated air is then drawn from behind the absorber plate into the building's ventilation system.

A Trombe wall is a passive solar heating and ventilation system consisting of an air channel sandwiched between a window and a sun-facing thermal mass. During the ventilation cycle, sunlight stores heat in the thermal mass and warms the air channel causing circulation through vents at the top and bottom of the wall. During the heating cycle the Trombe wall radiates stored heat.

Solar roof ponds are unique solar heating and cooling systems developed by Harold Hay in the 1960s. A basic system consists of a roof-mounted water bladder with a movable insulating cover. This system can control heat exchange between interior and exterior environments by covering and uncovering the bladder between night and day. When heating is a concern the bladder is uncovered during the day allowing sunlight to warm the water bladder and store heat for evening use. When cooling is a concern the covered bladder draws heat from the building's interior during the day and is uncovered at night to radiate heat to the cooler atmosphere. The Skytherm house in Atascadero, California uses a prototype roof pond for heating and cooling.

Solar space heating with solar air heat collectors is more popular in the USA and Canada than heating with solar liquid collectors since most buildings already have a ventilation system for heating and cooling. The two main types of solar air panels are glazed and unglazed.

Of the 21,000,000 square feet (2,000,000 m^2) of solar thermal collectors produced in the United States in 2007, 16,000,000 square feet (1,500,000 m^2) were of the low-temperature variety. Low-temperature collectors are generally installed to heat swimming pools, although they can also be used for space heating. Collectors can use air or water as the medium to transfer the heat to their destination.

Heat Storage in Low-temperature Solar Thermal Systems

Interseasonal storage: Solar heat (or heat from other sources) can be effectively stored between opposing seasons in aquifers, underground geological strata, large specially constructed pits, and large tanks that are insulated and covered with earth.

Short-term storage. Thermal mass materials store solar energy during the day and release this energy during cooler periods. Common thermal mass materials include stone, concrete, and water. The proportion and placement of thermal mass should consider several factors such as climate, daylighting, and shading conditions. When properly incorporated, thermal mass can passively maintain comfortable temperatures while reducing energy consumption.

Solar-driven Cooling

Worldwide, by 2011 there were about 750 cooling systems with solar-driven heat pumps, and annual market growth was 40 to 70% over the prior seven years. It is a niche market because the economics are challenging, with the annual number of cooling hours a limiting factor. Respectively, the annual cooling hours are roughly 1000 in the Mediterranean, 2500 in Southeast Asia, and only 50 to 200 in Central Europe. However, system construction costs dropped about 50% between 2007 and 2011. The International Energy Agency (IEA) Solar Heating and Cooling program (IEA-SHC) task groups working on further development of the technologies involved.

Solar Heat-driven Ventilation

A solar chimney (or thermal chimney) is a passive solar ventilation system composed of a hollow thermal mass connecting the interior and exterior of a building. As the chimney warms, the air inside is heated causing an updraft that pulls air through the building. These systems have been in use since Roman times and remain common in the Middle East.

Process Heat

Solar Evaporation Ponds in the Atacama Desert.

Solar process heating systems are designed to provide large quantities of hot water or space heating for nonresidential buildings.

Evaporation ponds are shallow ponds that concentrate dissolved solids through evaporation. The use of evaporation ponds to obtain salt from sea water is one of the oldest applications of solar energy. Modern uses include concentrating brine solutions used in leach mining and removing dissolved solids from waste streams. Altogether, evaporation ponds represent one of the largest commercial applications of solar energy in use today.

Unglazed transpired collectors are perforated sun-facing walls used for preheating ventilation air. Transpired collectors can also be roof mounted for year-round use and can raise the incoming air temperature up to 22 °C and deliver outlet temperatures of 45-60 °C. The short payback period of transpired collectors (3 to 12 years) make them a more cost-effective alternative to glazed collection systems. As of 2015, over 4000 systems with a combined collector area of 500,000 m² had been installed worldwide. Representatives include an 860 m² collector in Costa Rica used for drying coffee beans and a 1300 m² collector in Coimbatore, India used for drying marigolds.

A food processing facility in Modesto, California uses parabolic troughs to produce steam used in the manufacturing process. The 5,000 m² collector area is expected to provide 15 TJ per year.

Medium-temperature Collectors

These collectors could be used to produce approximately 50% and more of the hot water needed for residential and commercial use in the United States. In the United States, a typical system costs \$4000–\$6000 retail (\$1400 to \$2200 wholesale for the materials) and 30% of the system qualifies for a federal tax credit + additional state credit exists in about half of the states. Labor for a simple open loop system in southern climates can take 3–5 hours for the installation and 4–6 hours in Northern areas. Northern system require more collector area and more complex plumbing to protect the collector from freezing. With this incentive, the payback time for a typical household is four to nine years, depending on the state. Similar subsidies exist in parts of Europe. A crew of one solar plumber and two assistants with minimal training can install a system per day. Thermosiphon installation have negligible maintenance costs (costs rise if antifreeze and mains power are used for circulation) and in the US reduces a households' operating costs by \$6 per person per month. Solar water heating can reduce CO_2 emissions of a family of four by 1 ton/year (if replacing natural gas) or 3 ton/year (if replacing electricity). Medium-temperature installations can use any of several designs: common designs are pressurized glycol, drain back, batch systems and newer low pressure freeze tolerant systems using polymer pipes containing water with photovoltaic pumping. European and International standards are being reviewed to accommodate innovations in design and operation of medium temperature collectors. Operational innovations include "permanently wetted collector" operation. This innovation reduces or even eliminates the occurrence of no-flow high temperature stresses called stagnation which would otherwise reduce the life expectancy of collectors.

Solar Drying

Industrial indirect solar fruit and vegetable dryer.

Solar thermal energy can be useful for drying wood for construction and wood fuels such as wood chips for combustion. Solar is also used for food products such as fruits, grains, and fish. Crop drying by solar means is environmentally friendly as well as cost effective while improving the quality. The less money it takes to make a product, the less it can be sold for, pleasing both the buyers and the sellers. Technologies in solar drying include ultra low cost pumped transpired plate air collectors based on black fabrics. Solar thermal energy is helpful in the process of drying products such as wood chips and other forms of biomass by raising the temperature while allowing air to pass through and get rid of the moisture.

Cooking

The Solar Bowl above the Solar Kitchen in Auroville, India concentrates sunlight on a movable receiver to produce steam for cooking.

Solar cookers use sunlight for cooking, drying and pasteurization. Solar cooking offsets fuel costs, reduces demand for fuel or firewood, and improves air quality by reducing or removing a source of smoke.

The simplest type of solar cooker is the box cooker first built by Horace de Saussure in 1767. A basic box cooker consists of an insulated container with a transparent lid. These cookers can be used effectively with partially overcast skies and will typically reach temperatures of 50–100 °C.

Concentrating solar cookers use reflectors to concentrate solar energy onto a cooking container. The most common reflector geometries are flat plate, disc and parabolic trough type. These designs cook faster and at higher temperatures (up to 350 °C) but require direct light to function properly.

The Solar Kitchen in Auroville, India uses a unique concentrating technology known as the solar bowl. Contrary to conventional tracking reflector/fixed receiver systems, the solar bowl uses a fixed spherical reflector with a receiver which tracks the focus of light as the Sun moves across the sky. The solar bowl's receiver reaches temperature of 150 °C that is used to produce steam that helps cook 2,000 daily meals.

Many other solar kitchens in India use another unique concentrating technology known as the Scheffler reflector. This technology was first developed by Wolfgang Scheffler in 1986. A Scheffler reflector is a parabolic dish that uses single axis tracking to follow the Sun's daily course. These reflectors have a flexible reflective surface that is able to change its curvature to adjust to seasonal variations in the incident angle of sunlight. Scheffler reflectors have the advantage of having a fixed focal point which improves the ease of cooking and are able to reach temperatures of 450-650 °C. Built in 1999 by the Brahma Kumaris, the world's largest Scheffler reflector system in Abu Road, Rajasthan India is capable of cooking up to 35,000 meals a day. By early 2008, over 2000 large cookers of the Scheffler design had been built worldwide.

Distillation

Solar stills can be used to make drinking water in areas where clean water is not common. Solar distillation is necessary in these situations to provide people with purified water. Solar energy heats up the water in the still. The water then evaporates and condenses on the bottom of the covering glass.

High-temperature Collectors

Part of the 354 MW SEGS solar complex in northern San Bernardino County, California.

The solar furnace at Odeillo in the French Pyrenees-Orientales can reach temperatures up to 3,500°C .

Where temperatures below about 95 °C are sufficient, as for space heating, flat-plate collectors of the nonconcentrating type are generally used. Because of the relatively high heat losses through the glazing, flat plate collectors will not reach temperatures much above 200 °C even when the heat transfer fluid is stagnant. Such temperatures are too low for efficient conversion to electricity.

The efficiency of heat engines increases with the temperature of the heat source. To achieve this in solar thermal energy plants, solar radiation is concentrated by mirrors or lenses to obtain higher temperatures – a technique called Concentrated Solar Power (CSP). The practical effect of high efficiencies is to reduce the plant's collector size and total land use per unit power generated, reducing the environmental impacts of a power plant as well as its expense.

As the temperature increases, different forms of conversion become practical. Up to 600 °C, steam turbines, standard technology, have an efficiency up to 41%. Above 600 °C, gas turbines can be more efficient. Higher temperatures are problematic because different materials and techniques are needed. One proposal for very high temperatures is to use liquid fluoride salts operating between 700 °C to 800 °C, using multi-stage turbine systems to achieve 50% or more thermal efficiencies. The higher operating temperatures permit the plant to use higher-temperature dry heat exchangers for its thermal exhaust, reducing the plant's water use – critical in the deserts where large solar plants are practical. High temperatures also make heat storage more efficient, because more watt-hours are stored per unit of fluid.

Commercial concentrating solar thermal power (CSP) plants were first developed in the 1980s.

The world's largest solar thermal power plants are now the 370 MW Ivanpah Solar Power Facility, commissioned in 2014, and the 354 MW SEGS CSP installation, both located in the Mojave Desert of California, where several other solar projects have been realized as well. With the exception of the Shams solar power station, built in 2013 near Abu Dhabi, the United Arab Emirates, all other 100 MW or larger CSP plants are either located in the United States or in Spain.

The principal advantage of CSP is the ability to efficiently add thermal storage, allowing the dispatching of electricity over up to a 24-hour period. Since peak electricity demand typically occurs at about 5 pm, many CSP power plants use 3 to 5 hours of thermal storage. With current technology, storage of heat is much cheaper and more efficient than storage of electricity. In this way, the CSP plant can produce electricity day and night. If the CSP site has predictable solar radiation, then the CSP plant becomes a reliable power plant. Reliability can further be improved by installing a back-up combustion system. The back-up system can use most of the CSP plant, which decreases the cost of the back-up system.

CSP facilities utilize high electrical conductivity materials, such as copper, in field power cables, grounding networks, and motors for tracking and pumping fluids, as well as in the main generator and high voltage transformers.

With reliability, unused desert, no pollution, and no fuel costs, the obstacles for large deployment for CSP are cost, aesthetics, land use and similar factors for the necessary connecting high tension lines. Although only a small percentage of the desert is necessary to meet global electricity demand, still a large area must be covered with mirrors or lenses to obtain a significant amount of energy. An important way to decrease cost is the use of a simple design.

When considering land use impacts associated with the exploration and extraction through to transportation and conversion of fossil fuels, which are used for most of our electrical power, utility-scale solar power compares as one of the most land-efficient energy resources available:

The federal government has dedicated nearly 2,000 times more acreage to oil and gas leases than to solar development. In 2010 the Bureau of Land Management approved nine large-scale solar projects, with a total generating capacity of 3,682 megawatts, representing approximately 40,000 acres. In contrast, in 2010, the Bureau of Land Management processed more than 5,200 applications gas and oil leases, and issued 1,308 leases, for a total of 3.2 million acres. Currently, 38.2 million acres of onshore public lands and an additional 36.9 million acres of offshore exploration in the Gulf of Mexico are under lease for oil and gas development, exploration and production.

System Designs

During the day the sun has different positions. For low concentration systems (and low temperatures) tracking can be avoided (or limited to a few positions per year) if nonimaging optics are used. For higher concentrations, however, if the mirrors or lenses do not move, then the focus of the mirrors or lenses changes (but also in these cases nonimaging optics provides the widest acceptance angles for a given concentration). Therefore, it seems unavoidable that there needs to be a tracking system that follows the position of the sun (for solar photovoltaic a solar tracker is only optional). The tracking system increases the cost and complexity. With this in mind, different designs can be distinguished in how they concentrate the light and track the position of the sun.

Parabolic Trough Designs

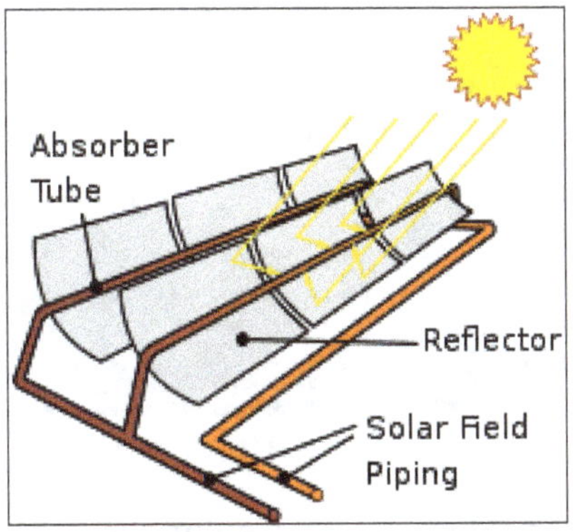

Sketch of a parabolic trough design. A change of position of the sun parallel to the receiver does not require adjustment of the mirrors.

Parabolic trough power plants use a curved, mirrored trough which reflects the direct solar radiation onto a glass tube containing a fluid (also called a receiver, absorber or collector) running the length of the trough, positioned at the focal point of the reflectors. The trough is parabolic along one axis and linear in the orthogonal axis. For change of the daily position of the sun perpendicular to the receiver, the trough tilts east to west so that the direct radiation remains focused on the receiver. However, seasonal changes in the angle of sunlight parallel to the trough does not require adjustment of the mirrors, since the light is simply concentrated elsewhere on the receiver. Thus the trough design does not require tracking on a second axis. The receiver may be enclosed in a glass vacuum chamber. The vacuum significantly reduces convective heat loss.

A fluid (also called heat transfer fluid) passes through the receiver and becomes very hot. Common fluids are synthetic oil, molten salt and pressurized steam. The fluid containing the heat is transported to a heat engine where about a third of the heat is converted to electricity.

Full-scale parabolic trough systems consist of many such troughs laid out in parallel over a large area of land. Since 1985 a solar thermal system using this principle has been in full operation in California in the United States. It is called the Solar Energy Generating Systems (SEGS) system. Other CSP designs lack this kind of long experience and therefore it can currently be said that the parabolic trough design is the most thoroughly proven CSP technology.

The SEGS is a collection of nine plants with a total capacity of 354 MW and has been the world's largest solar power plant, both thermal and non-thermal, for many years. A newer plant is Nevada Solar One plant with a capacity of 64 MW. The 150 MW Andasol solar power stations are in Spain with each site having a capacity of 50 MW. Note however, that those plants have heat storage which requires a larger field of solar collectors relative to the size of the steam turbine-generator to store heat and send heat to the steam turbine at the same time. Heat storage enables better utilization of the steam turbine. With day and some nighttime operation of the steam-turbine Andasol 1 at 50 MW peak capacity produces more energy than Nevada Solar One at 64 MW peak capacity, due to the former plant's thermal energy storage system and larger solar field. The 280MW Solana Generating Station came online in Arizona in 2013 with 6 hours of power storage. Hassi R'Mel integrated solar combined cycle power station in Algeria and Martin Next Generation Solar Energy Center both use parabolic troughs in a combined cycle with natural gas.

Enclosed Trough

Inside an enclosed trough system

The enclosed trough architecture encapsulates the solar thermal system within a greenhouse-like glasshouse. The glasshouse creates a protected environment to withstand the elements that can negatively impact reliability and efficiency of the solar thermal system.

Lightweight curved solar-reflecting mirrors are suspended within the glasshouse structure. A single-axis tracking system positions the mirrors to track the sun and focus its light onto a network of stationary steel pipes, also suspended from the glasshouse structure. Steam is generated directly, using oil field-quality water, as water flows from the inlet throughout the length of the pipes, without heat exchangers or intermediate working fluids.

The steam produced is then fed directly to the field's existing steam distribution network, where the steam is continuously injected deep into the oil reservoir. Sheltering the mirrors from the wind allows them to achieve higher temperature rates and prevents dust from building up as a result from exposure to humidity. GlassPoint Solar, the company that created the Enclosed Trough design, states its technology can produce heat for EOR for about $5 per million British thermal units in sunny regions, compared to between $10 and $12 for other conventional solar thermal technologies.

Power Tower Designs

Ivanpah Solar Electric Generating System with all three towers under load, Feb., 2014. Taken from I-15 in San Bernardino County, California. The Clark Mountain Range can be seen in the distance.

Power towers (also known as 'central tower' power plants or 'heliostat' power plants) capture and focus the sun's thermal energy with thousands of tracking mirrors (called heliostats) in roughly a two square mile field. A tower resides in the center of the heliostat field. The heliostats focus concentrated sunlight on a receiver which sits on top of the tower. Within the receiver the concentrated sunlight heats molten salt to over 1,000 °F (538 °C). The heated molten salt then flows into a thermal storage tank where it is stored, maintaining 98% thermal efficiency, and eventually pumped to a steam generator. The steam drives a standard turbine to generate electricity. This process, also known as the "Rankine cycle" is similar to a standard coal-fired power plant, except it is fueled by clean and free solar energy.

The advantage of this design above the parabolic trough design is the higher temperature. Thermal energy at higher temperatures can be converted to electricity more efficiently and can be more cheaply stored for later use. Furthermore, there is less need to flatten the ground area. In principle a power tower can be built on the side of a hill. Mirrors can be flat and plumbing is concentrated in the tower. The disadvantage is that each mirror must have its own dual-axis control, while in the parabolic trough design single axis tracking can be shared for a large array of mirrors.

A cost/performance comparison between power tower and parabolic trough concentrators was made by the NREL which estimated that by 2020 electricity could be produced from power towers for 5.47 ¢/kWh and for 6.21 ¢/kWh from parabolic troughs. The capacity factor for power towers was estimated to be 72.9% and 56.2% for parabolic troughs. There is some hope that the development of cheap, durable, mass producible heliostat power plant components could bring this cost down.

The first commercial tower power plant was PS10 in Spain with a capacity of 11 MW, completed in 2007. Since then a number of plants have been proposed, several have been built on a number of countries (Spain, Germany, U.S., Turkey, China, India) but several proposed plants were cancelled as photovoltaic solar prices plummeted. A solar power tower is expected to come online in South Africa in 2014. Ivanpah Solar Power Facility in California generates 392 MW of electricity from three towers, making it the largest solar power tower plant when it came online in late 2013.

Dish Designs

A parabolic solar dish concentrating the sun's rays on the heating element of a Stirling engine. The entire unit acts as a solar tracker.

CSP-Stirling is known to have the highest efficiency of all solar technologies (around 30%, compared to solar photovoltaic's approximately 15%), and is predicted to be able to produce the cheapest energy among all renewable energy sources in high-scale production and hot areas, semi-deserts, etc. A dish Stirling system uses a large, reflective, parabolic dish (similar in shape to a satellite television dish). It focuses all the sunlight that strikes the dish up onto a single point above the dish, where a receiver captures the heat and transforms it into a useful form. Typically the dish is coupled with a Stirling engine in a Dish-Stirling System, but also sometimes a steam engine is used. These create rotational kinetic energy that can be converted to electricity using an electric generator.

In 2005 Southern California Edison announced an agreement to purchase solar powered Stirling engines from Stirling Energy Systems over a twenty-year period and in quantities (20,000 units) sufficient to generate 500 megawatts of electricity. In January 2010, Stirling Energy Systems and Tessera Solar commissioned the first demonstration 1.5-megawatt power plant ("Maricopa Solar") using Stirling technology in Peoria, Arizona. At the beginning of 2011 Stirling Energy's development arm, Tessera Solar, sold off its two large projects, the 709 MW Imperial project and the 850 MW Calico project to AES Solar and K.Road, respectively. In 2012 the Maricopa plant was bought and dismantled by United Sun Systems. United Sun Systems released a new generation system, based on a V-shaped Stirling engine and a peak production of 33 kW. The new CSP-Stirling technology brings down LCOE to USD 0.02 in utility scale.

According to its developer, Rispasso Energy, a Swedish firm, in 2015 its Dish Sterling system being tested in the Kalahari Desert in South Africa showed 34% efficiency.

Fresnel Technologies

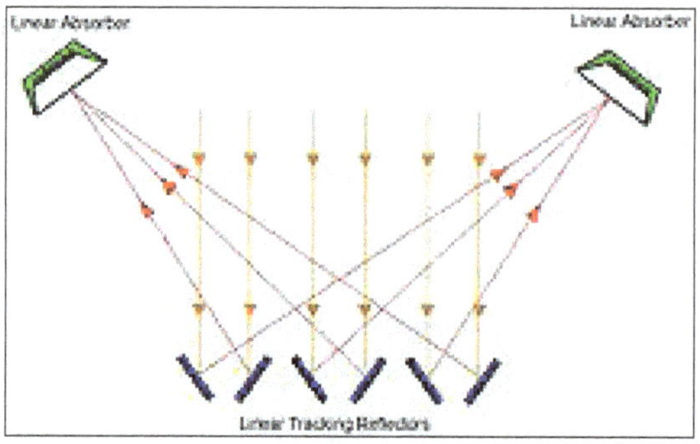

Fresnel reflector.

A linear Fresnel reflector power plant uses a series of long, narrow, shallow-curvature (or even flat) mirrors to focus light onto one or more linear receivers positioned above the mirrors. On top of the receiver a small parabolic mirror can be attached for further focusing the light. These systems aim to offer lower overall costs by sharing a receiver between several mirrors (as compared with trough and dish concepts), while still using the simple line-focus geometry with one axis for tracking. This is similar to the trough design (and different from central towers and dishes with dual-axis). The receiver is stationary and so fluid couplings are not required (as in troughs and dishes). The mirrors also do not need to support the receiver, so they are structurally simpler. When suitable aiming strategies are used (mirrors aimed at different receivers at different times of day), this can allow a denser packing of mirrors on available land area.

Rival single axis tracking technologies include the relatively new linear Fresnel reflector (LFR) and compact-LFR (CLFR) technologies. The LFR differs from that of the parabolic trough in that the absorber is fixed in space above the mirror field. Also, the reflector is composed of many low row segments, which focus collectively on an elevated long tower receiver running parallel to the reflector rotational axis.

Prototypes of Fresnel lens concentrators have been produced for the collection of thermal energy by International Automated Systems. No full-scale thermal systems using Fresnel lenses are known to be in operation, although products incorporating Fresnel lenses in conjunction with photovoltaic cells are already available.

MicroCSP

MicroCSP is used for community-sized power plants (1 MW to 50 MW), for industrial, agricultural and manufacturing 'process heat' applications, and when large amounts of hot water are needed, such as resort swimming pools, water parks, large laundry facilities, sterilization, distillation and other such uses.

Enclosed Parabolic Trough

The enclosed parabolic trough solar thermal system encapsulates the components within an off-the-shelf greenhouse type of glasshouse. The glasshouse protects the components from the elements that can negatively impact system reliability and efficiency. This protection importantly includes nightly glass-roof washing with optimized water-efficient off-the-shelf automated washing systems. Lightweight curved solar-reflecting mirrors are suspended from the ceiling of the glasshouse by wires. A single-axis tracking system positions the mirrors to retrieve the optimal amount of sunlight. The mirrors concentrate the sunlight and focus it on a network of stationary steel pipes, also suspended from the glasshouse structure. Water is pumped through the pipes and boiled to generate steam when intense sun radiation is applied. The steam is available for process heat. Sheltering the mirrors from the wind allows them to achieve higher temperature rates and prevents dust from building up on the mirrors as a result from exposure to humidity.

Heat Collection and Exchange

More energy is contained in higher frequency light based upon the formula of $E = h\nu$, where h is the Planck constant and ν is frequency. Metal collectors down convert higher frequency light by producing a series of Compton shifts into an abundance of lower frequency light. Glass or ceramic coatings with high transmission in the visible and UV and effective absorption in the IR (heat blocking) trap metal absorbed low frequency light from radiation loss. Convection insulation prevents mechanical losses transferred through gas. Once collected as heat, thermos containment efficiency improves significantly with increased size. Unlike Photovoltaic technologies that often degrade under concentrated light, Solar Thermal depends upon light concentration that requires a clear sky to reach suitable temperatures.

Heat in a solar thermal system is guided by five basic principles: heat gain; heat transfer; heat storage; heat transport; and heat insulation. Here, heat is the measure of the amount of thermal energy an object contains and is determined by the temperature, mass and specific heat of the object. Solar thermal power plants use heat exchangers that are designed for constant working conditions, to provide heat exchange. Copper heat exchangers are important in solar thermal heating and cooling systems because of copper's high thermal conductivity, resistance to atmospheric and water corrosion, sealing and joining by soldering, and mechanical strength. Copper is used both in receivers and in primary circuits (pipes and heat exchangers for water tanks) of solar thermal water systems.

Heat gain is the heat accumulated from the sun in the system. Solar thermal heat is trapped using the greenhouse effect; the greenhouse effect in this case is the ability of a reflective surface to transmit short wave radiation and reflect long wave radiation. Heat and infrared radiation (IR) are produced when short wave radiation light hits the absorber plate, which is then trapped inside the collector. Fluid, usually water, in the absorber tubes collect the trapped heat and transfer it to a heat storage vault.

Heat is transferred either by conduction or convection. When water is heated, kinetic energy is transferred by conduction to water molecules throughout the medium. These molecules spread their thermal energy by conduction and occupy more space than the cold slow moving molecules above them. The distribution of energy from the rising hot water to the sinking cold water contributes to the convection process. Heat is transferred from the absorber plates of the collector in the fluid by conduction. The collector fluid is circulated through the carrier pipes to the heat transfer vault. Inside the vault, heat is transferred throughout the medium through convection.

Heat storage enables solar thermal plants to produce electricity during hours without sunlight. Heat is transferred to a thermal storage medium in an insulated reservoir during hours with sunlight, and is withdrawn for power generation during hours lacking sunlight. Thermal storage mediums will be discussed in a heat storage section. Rate of heat transfer is related to the conductive and convection medium as well as the temperature differences. Bodies with large temperature differences transfer heat faster than bodies with lower temperature differences.

Heat transport refers to the activity in which heat from a solar collector is transported to the heat storage vault. Heat insulation is vital in both heat transport tubing as well as the storage vault. It prevents heat loss, which in turn relates to energy loss, or decrease in the efficiency of the system.

Heat Storage for Space Heating

A collection of mature technologies called seasonal thermal energy storage (STES) is capable of storing heat for months at a time, so solar heat collected primarily in Summer can be used for all-year heating. Solar-supplied STES technology has been advanced primarily in Denmark, Germany, and Canada, and applications include individual buildings and district heating networks. Drake Landing Solar Community in Alberta, Canada has a small district system and in 2012 achieved a world record of providing 97% of the community's all-year space heating needs from the sun. STES thermal storage mediums include deep aquifers; native rock surrounding clusters of small-diameter, heat exchanger equipped boreholes; large, shallow, lined pits that are filled with gravel and top-insulated; and large, insulated and buried surface water tanks.

Heat Storage to Stabilize Solar-electric Power Generation

Heat storage allows a solar thermal plant to produce electricity at night and on overcast days. This allows the use of solar power for baseload generation as well as peak power generation, with the potential of displacing both coal- and natural gas-fired power plants. Additionally, the utilization of the generator is higher which reduces cost. Even short term storage can help by smoothing out the "duck curve" of rapid change in generation requirements at sunset when a grid includes large amounts of solar capacity.

Heat is transferred to a thermal storage medium in an insulated reservoir during the day, and withdrawn for power generation at night. Thermal storage media include pressurized steam, concrete, a variety of phase change materials, and molten salts such as calcium, sodium and potassium nitrate.

Steam Accumulator

The PS10 solar power tower stores heat in tanks as pressurized steam at 50 bar and 285 °C. The steam condenses and flashes back to steam, when pressure is lowered. Storage is for one hour. It is suggested that longer storage is possible, but that has not been proven yet in an existing power plant.

Molten Salt Storage

The 150 MW Andasol solar power station is a commercial parabolic trough solar thermal power plant, located in Spain. The Andasol plant uses tanks of molten salt to store solar energy so that it can continue generating electricity even when the sun isn't shining.

A variety of fluids have been tested to transport the sun's heat, including water, air, oil, and sodium, but Rockwell International selected molten salt as best. Molten salt is used in solar power tower systems because it is liquid at atmospheric pressure, provides a low-cost medium to store thermal energy, its operating temperatures are compatible with today's steam turbines, and it is non-flammable and nontoxic. Molten salt is used in the chemical and metals industries to transport heat, so industry has experience with it.

The first commercial molten salt mixture was a common form of saltpeter, 60% sodium nitrate and 40% potassium nitrate. Saltpeter melts at 220 °C (430 °F) and is kept liquid at 290 °C (550 °F) in an insulated storage tank. Calcium nitrate can reduce the melting point to 131 °C, permitting more energy to be extracted before the salt freezes. There are now several technical calcium nitrate grades stable at more than 500 °C.

This solar power system can generate power in cloudy weather or at night using the heat in the tank of hot salt. The tanks are insulated, able to store heat for a week. Tanks that power a 100-megawatt turbine for four hours would be about 9 m (30 ft) tall and 24 m (80 ft) in diameter.

The Andasol power plant in Spain is the first commercial solar thermal power plant using molten salt for heat storage and nighttime generation. It came on line March 2009. On July 4, 2011, a company in Spain celebrated an historic moment for the solar industry: Torresol's 19.9 MW concentrating solar power plant became the first ever to generate uninterrupted electricity for 24 hours straight, using a molten salt heat storage.

In 2016 SolarReserve proposed a 2 GW, $5 billion concentrated solar plant with storage in Nevada.

Phase-change Materials for Storage

Phase Change Material (PCMs) offer an alternative solution in energy storage. Using a similar heat transfer infrastructure, PCMs have the potential of providing a more efficient means of storage. PCMs can be either organic or inorganic materials. Advantages of organic PCMs include no corrosives, low or no undercooling, and chemical and thermal stability. Disadvantages include low phase-change enthalpy, low thermal conductivity, and flammability. Inorganics are advantageous with greater phase-change enthalpy, but exhibit disadvantages with undercooling, corrosion, phase separation, and lack of thermal stability. The greater phase-change enthalpy in inorganic PCMs make hydrate salts a strong candidate in the solar energy storage field.

Use of Water

A design which requires water for condensation or cooling may conflict with location of solar thermal plants in desert areas with good solar radiation but limited water resources. The conflict is illustrated by plans of Solar Millennium, a German company, to build a plant in the Amargosa Valley of Nevada which would require 20% of the water available in the area. Some other projected plants by the same and other companies in the Mojave Desert of California may also be affected by difficulty in obtaining adequate and appropriate water rights. California water law currently prohibits use of potable water for cooling.

Other designs require less water. The Ivanpah Solar Power Facility in south-eastern California conserves scarce desert water by using air-cooling to convert the steam back into water. Compared to conventional wet-cooling, this results in a 90% reduction in water usage at the cost of some loss of efficiency. The water is then returned to the boiler in a closed process which is environmentally friendly.

Conversion Rates from Solar Energy to Electrical Energy

Of all of these technologies the solar dish/Stirling engine has the highest energy efficiency. A single solar dish-Stirling engine installed at Sandia National Laboratories National Solar Thermal Test Facility (NSTTF) produces as much as 25 kW of electricity, with a conversion efficiency of 31.25%.

Solar parabolic trough plants have been built with efficiencies of about 20%. Fresnel reflectors have an efficiency that is slightly lower (but this is compensated by the denser packing).

The gross conversion efficiencies (taking into account that the solar dishes or troughs occupy only a fraction of the total area of the power plant) are determined by net generating capacity over the solar energy that falls on the total area of the solar plant. The 500-megawatt (MW) SCE/SES plant would extract about 2.75% of the radiation that falls on its 4,500 acres (18.2 km^2). For the 50 MW AndaSol Power Plant that is being built in Spain (total area of 1,300×1,500 m = 1.95 km^2) gross conversion efficiency comes out at 2.6%.

Furthermore, efficiency does not directly relate to cost: on calculating total cost, both efficiency and the cost of construction and maintenance should be taken into account.

Solar Thermal Power Plants

The two main types of solar thermal power plants are

1. Concentrating Solar Power (CSP) plants.
2. Solar Chimneys.

Concentrating Solar Power (CSP) Plants

Solar thermal power plants generally use reflectors to concentrate sunlight into a heat absorber. Such power plants are known as Concentrating Solar Power (CSP) plants.

Concentrating solar power plants produce electric power by converting the sun's energy into high-temperature heat using various mirror configurations. The heat is then channeled through a conventional generator. The plants consist of two parts, one that collects solar energy and converts it to heat, and another that converts heat energy to electricity.

Concentrating solar power systems can be sized for village power (10 kilowatts) or grid- connected applications (up to 100 megawatts). Some systems use thermal storage during cloudy periods or at night. There are four CSP technologies being promoted internationally. For each of these, there exists various design variations or different configurations.

The amount of power generated by a concentrating solar power plant depends on the amount of direct sunlight. Like concentrating photovoltaic concentrators, these technologies use only direct-beam sunlight, rather than diffuse solar radiation.

Types of CSP plants

A. Parabolic Trough Systems:

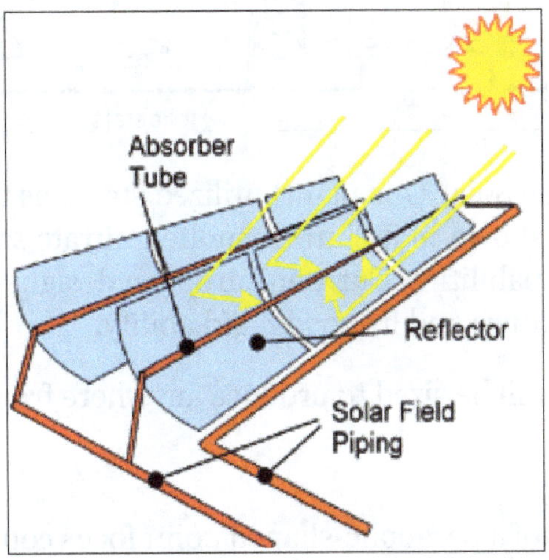

The sun's energy is concentrated by parabolically curved, trough-shaped reflectors onto a receiver pipe running along the inside of the curved surface. This energy heats oil flowing through the pipe and the heat energy is then used to generate electricity in a conventional steam generator.

A collector field comprises many troughs in parallel rows aligned on a north-south axis. This configuration enables the single-axis troughs to track the sun from east to west during the day to ensure that the sun is continuously focused on the receiver pipes. Individual trough systems currently can generate about 80 megawatts of electricity.

Trough designs can incorporate thermal storage—setting aside the heat transfer fluid in its hot phase—allowing for electricity generation several hours into the evening.

Currently, all parabolic trough plants are "hybrids," meaning they use fossil fuel to supplement the solar output during periods of low solar radiation.

Another option under investigation is the approximation of the parabolic troughs by segmented mirrors according to the principle of Fresnel.

B. Power Tower Systems:

A power tower converts sunshine into clean electricity for the electricity grids. The technology utilizes many large, sun-tracking mirrors (heliostats) to focus sunlight on a receiver at the top of a tower. A heat transfer fluid heated in the receiver is used to generate steam, which, in turn, is used in a conventional turbine-generator to produce electricity.

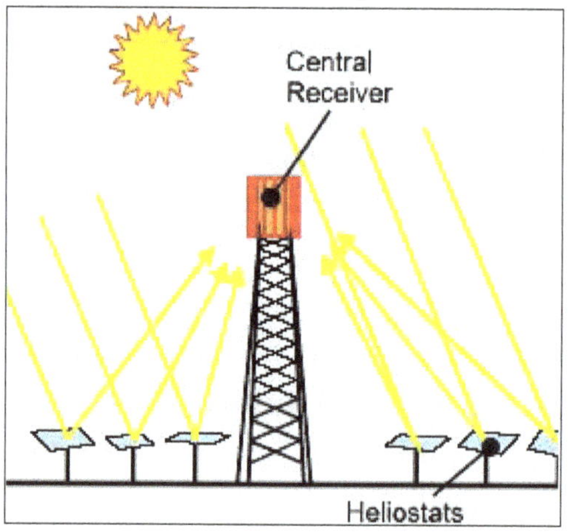

Early power towers (such as the Solar One plant) utilized steam as the heat transfer fluid; current designs (including Solar Two, shown in fig) utilize molten nitrate salt because of its superior heat transfer and energy storage capabilities. Current European designs use air as heat transfer medium because of its high temperature and its good hand ability.

Individual commercial plants will be sized to produce anywhere from 50 to 200 MW of electricity.

C. Parabolic Dish Systems:

Parabolic dish systems consist of a parabolic-shaped point focus concentrator in the form of a dish that reflects solar radiation onto a receiver mounted at the focal point. These concentrators are mounted on a structure with a two-axis tracking system to follow the sun.

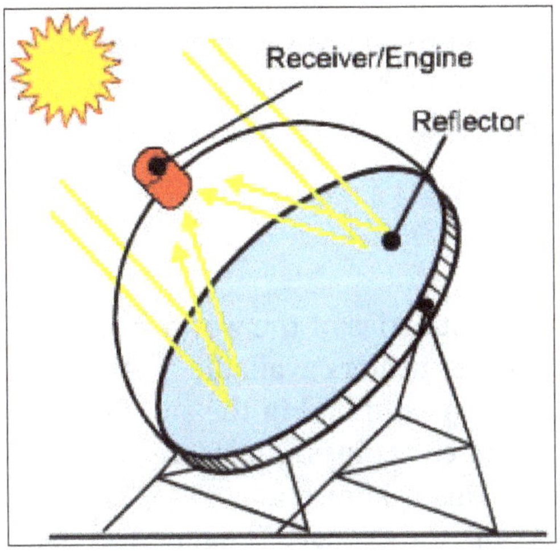

The collected heat is typically utilized directly by a heat engine mounted on the receiver moving with the dish structure. Stirling and Brayton cycle engines are currently favored for power conversion.

Projects of modular systems have been realized with total capacities up to 5 MW. The modules have max sizes of 50 kW and have achieved peak efficiencies up to 30% net.

Solar Chimney

A solar chimney is a solar thermal power plant where air passes under a very large agricultural glass house (between 2 and 30 kilometers in diameter); the air is heated by the sun and channeled upwards towards a convection tower. It then rises naturally and is used to drive turbines, which generate electricity.

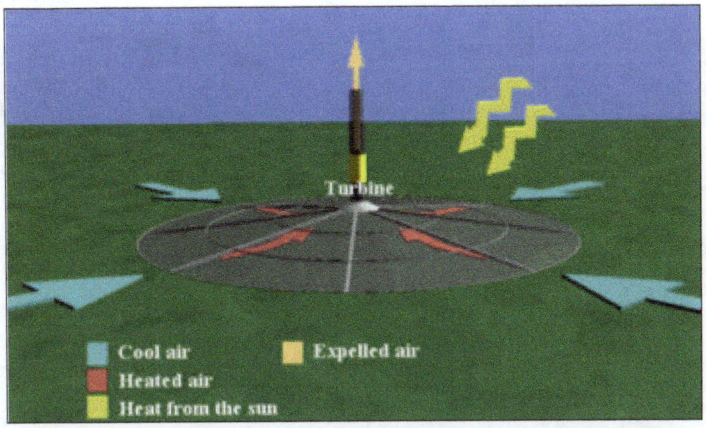

General concept of proposed solar chimney power station.

A solar chimney is an days such a chimney can provide ventilation where there apparatus for harnessing solar energy by convection of heated air. In its simplest form, it simply consists of a black-painted chimney. During the daytime, solar energy heats the chimney and thereby heats the air within it, resulting in an Updraft of air within the chimney. The suction this creates at the chimney base can also be used to ventilate, and thereby cool, the building below. In most parts of the world, it is easier to harness wind power for such ventilation, but on hot windless would otherwise be none. This principle has been proposed for electric power generation, using a large greenhouse at the base rather than relying on heating of the chimney itself. The main problem with this approach is the relatively small difference in temperature between the highest and lowest temperatures in the system. Carnot's theorem greatly restricts the efficiency of conversion in these circumstances.

Water Heating

Water heating is required in most countries of the world for both domestic and commercial use. There are a wide variety of solar water heaters available. The simplest solar water heater is a piece of black plastic pipe, filled with water, and laid in the sun for the water to heat up. Simple solar water heaters usually comprise a series of pipes, which are painted black, sitting inside an insulated box fronted with a glass panel. This is known as a solar collector. The fluid to be heated passes through the collector and into a tank for storage. The fluid can be cycled through the tank several times to raise the heat of the fluid to the required temperature. There are two common simple configurations for such a system and they are outlined below.

a. The thermosyphon system makes use of the natural tendency of hot water to rise above cold water. The tank in such a system is always placed above the top of the collector and as water is heated in the collector it rises and is replaced by cold water from the bottom of the tank. This cycle will continue until the temperature of the water in the tank is equal to that of the panel. A one-way valve is usually fitted in the system to prevent the reverse occurring at night when the temperature drops. As hot water is drawn off for use, fresh cold water is fed into the system from the mains. As most solar collectors are fitted on the roofs of houses, this system is not always convenient, as it is difficult to site the tank above the collector, in which case the system will need A pump to circulate the water.

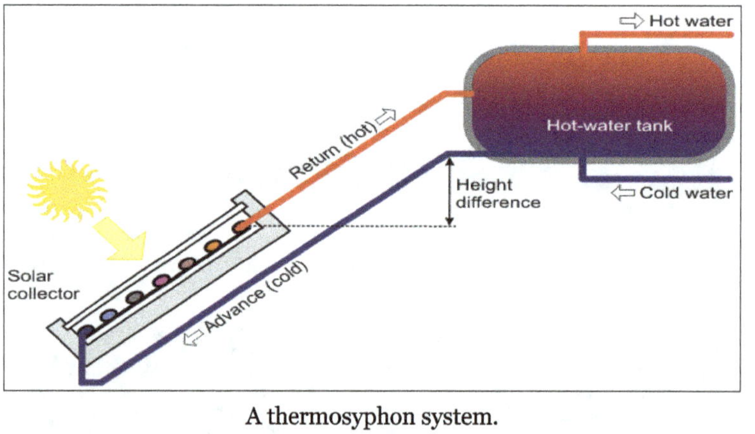

A thermosyphon system.

b. Pumped solar water heaters use a pumping device to drive the water through the collector. The advantage of this system is that the storage tank can be sited below the collector. The disadvantage of course is that electricity is required to drive the pump. Often the fluid circulating in the collector will be treated with an anti-corrosive and /or anti-freeze chemical. In this case a heat exchanger is required to transfer the heat to the consumers hot water supply.

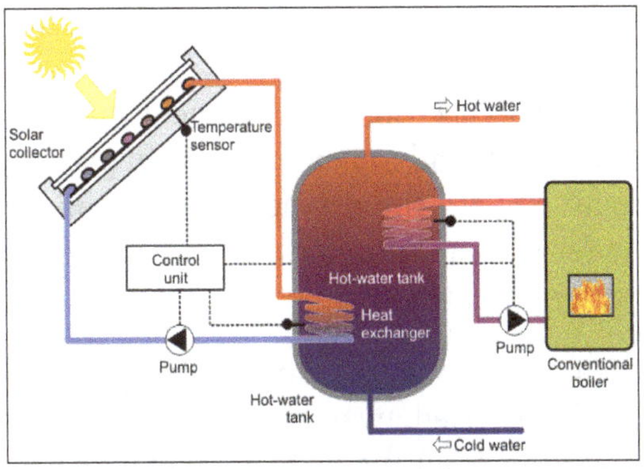

A double-cycle system with forced circulation with a conventional boiler for back-up heating.

Solar District Heating

If an entire housing estate should be fitted with solar systems, one solution is a solar district heating system. The collectors are either distributed on the houses, or replaced by a large, central solar collector. The collectors then heat up a big central storage tank, from which much of the heat is distributed back to the houses. The surface-to-volume ratio of a central storage tank is much better than that for distributed storage systems, so the storage losses are much lower, and even permit seasonal heat storage. Solar district heating is also an option if room heating is to be covered by solar energy. There are higher piping losses with a central tank, but some solar district heating demonstration systems have already been successfully tested.

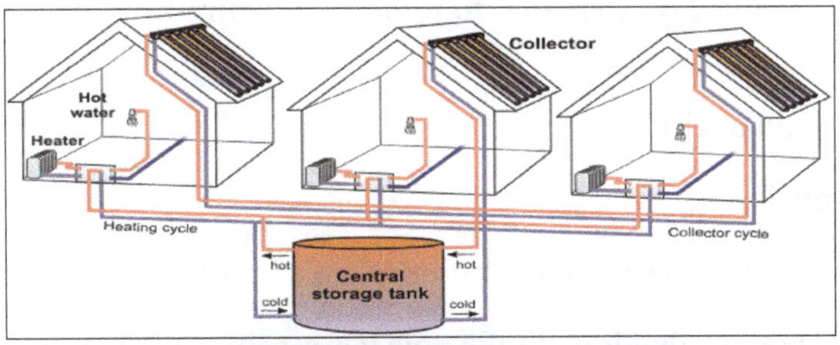

A solar district heating system.

Cost Benefits of Solar Water Heating System

The most cost- effective way to install a solar geyser is to integrate the collector assembly, cold-water supply and piping with the design of a new house under construction. Solar geysers can easily be installed in group houses and apartments, especially during construction, if adequate provisions are made for piping, collector assembly and cold-water supply. Proper load matching is required to ensure that the capacity of the system installed is optimized to meet The daily hot water needs of the end-user.

Current prices of domestic SWHs are around Rest. 20,000 for a 100 litres per day system. The life-cycle cost of energy of saved electrical energy is Rest. 5600 per year at the point of end- user. If the SWH is purchased as an add-on to an existing geyser, the simple payback period for the consumer would be 5.45 years. If the SWH is purchased instead of a conventional geyser, the payback period for the consumer would reduce to 3.58 years.

Solar Dryer

Controlled drying is required for various crops and products, such as grain, coffee, tobacco, fruits vegetables and fish. Their quality can be enhanced if the drying is properly carried out. Solar thermal technology can be used to assist with the drying of such products. Solar drying is in practice since the time imp-memorable for preservation of food and agriculture crops. This was done particularly by open sun drying under open the sky.

In open air Solar drying the heat is supplied by direct absorption of solar radiation by material being dried. The vapor produced is carried away by air moving past the material, the air motion being due either to natural convection resulting from contact with the heated material or to winds.

This process has several disadvantages.

Disadvantages of Mechanical and Artificial Drying:

1. Spoilage of product due to adverse climatic condition like rain, wind etc.
2. Loss of material due to birds and animals.
3. Deterioration of the material by decomposition, insects and fungus growth.
4. Highly energy intensive and expensive.

Solar dryer make use of solar radiation, ambient temperature, relative humidity. Heated air is passed naturally or mechanically circulated to remove moisture from material placed in side the enclosure.

Solar dryer is a very useful device for:

1. Agriculture crop drying.
2. Food processing industries for dehydration of fruits, potatoes, onions and other vegetables,
3. Dairy industries for production of milk powder, casein etc.
4. Seasoning of wood and timber.
5. Textile industries for drying of textile materials.

Working of Solar Dryer

The main principle of operation is to raise the heat of the product, which is usually held within a compartment or box, while at the same time passing air through the compartment to remove moisture. The flow of air is often promoted using the 'stack' effect which takes advantage of the fact that hot air rises and can therefore be drawn upwards through a chimney, while drawing in Cooler air from below. Alternatively a fan can be used. The size and shape of the compartment Varies depending on the product and the scale of the drying system. Large systems can use Large barns while smaller systems may have a few trays in a small wooden housing.

Solar crop drying technologies can help reduce environmental degradation caused by the use Of fuel wood or fossil fuels for crop drying and can also help to reduce the costs associated.

With these fuels and hence the cost of the product.

The principal types of solar dryers are enumerated below.

1. Solar cabinet dryer.
2. Solar green house dryers.
3. Indirects sonars drayer.

Solar Stills

Solar still is a device to desalinate impure water like brackish or saline water. It a simple device to get potable/fresh distilled water from impure water, using solar energy as fuel, for its various applications in domestic, industrial and academic sectors. A solar still consist of shallow triangular basin made up of Fiber Reinforced Plastic (FRP). Bottom of the basin is painted black so as to absorb solar heat effectively. Top of the basin is covered with transparent glass tilt fitted so that maximum solar radiation can be transmitted in to the still. Edges of the glass are sealed with the basin using tar tape so that the entire basin becomes airtight. Entire assembly is placed on a structure made of MS angle. Out let is connected with a storage container. Provision has been made to fill water in the still basin. A window is provided in the basin to clean the basin from inside. Water is charged in to the basin in a thin layer.

Solar Stills have got major advantages over other conventional Distillation / water purification / de- mineralization systems as follows:

1. Produces pure water.
2. No prime movers required.
3. No conventional energy required.
4. No skilled operator required.
5. Local manufacturing/repairing.
6. Low investment.
7. Can purify highly saline water (even sea water).

Working of Solar Still

Working of solar still is based on simple scientific principle of Evaporation and condensation. Brackish or saline water is filled in still basin, which is painted black at the bottom. Solar radiation received at the surface is absorbed effectively by the blacken surface and heat is transferred to the water in the basin. Temperature of the water increases that increases rate of evaporation. Water vapour formed by evaporation rises upward and condenses on the inner surface of the cover glass, which is relatively cold.

Condensed water vapor trickles down in to troughs from there it is collected in to the storage container.

The operation of Solar Still is very simple and no special skill is required for its operation and maintenance. Solar stills is an useful devise to get fresh/ distilled water which is required in Industries for industrial processes Hospitals and Dispensaries for sterilization Garages and Automobile Workshop for radiator and battery maintenance Telephone Exchange for battery maintenance Laboratory Use for analytic work Marshy and costal area To get fresh potable water.

Solar Box Cooker

A solar box cooker is an insulated transparent-topped box with a reflective lid. It is designed to capture solar power and keep its interior warm. The major parts of a solar cooker are enumerated below.

Important Parts of Solar Cooker

1. Outer Box: The outer box of a solar cooker is generally made of G.I. or aluminum sheet or fiber reinforced plastic.
2. Inner Cooking Box (Tray): This is made from aluminum sheet. The inner cooking box is slightly smaller than the outer box. It is coated with black paint so as to easily absorb solar radiation and transfer the heat to the cooking pots.
3. Double Glass Lid: A double glass lid covers the inner box or tray. This cover is slightly larger than the inner box. The two glass sheets are fixed in an aluminum frame with a spacing of 2 centimeters between the two glasses. This space contains air which insulates and prevents heat escaping from inside. A rubber strip is affixed on the edges of the frame to prevent any heat leakage.
4. Thermal Insulator: The space between the outer box and inner tray including bottom of the tray is packed with insulating material such as glass wool pads to reduce heat losses from the cooker. This insulating material should be free from volatile materials.
5. Mirror: Mirror is used in a solar cooker to increase the radiation input on the absorbing space and is fixed on the inner side of the main cover of the box. Sunlight falling on the mirror gets reflected from it and enters into the tray through the double glass lid. This radiation is in addition to the radiation entering the box directly and helps to quicken the cooking process by raising the inside temperature of the cooker.

6. Containers: The cooking containers (with cover) are generally made of aluminum or stainless steel. These pots are also painted black on the outer surface so that they also absorb solar radiation directly.

The solar box cooker typically reaches a temperature of 90 °C; not as hot as a standard oven, but still enough to warm food over an hour. Because it doesn't reach too high a temperature, food can be safely left in it all day without burning. The cooker is often used to make a large pot of food in the morning, and then people eat servings or snack from it all day. The cooker is usually used to warm food and drinks but can also be used to pasteurize milk and sanitize water.

Horace de Assure, a Swiss naturalist, invented solar cookers as early as 1767.

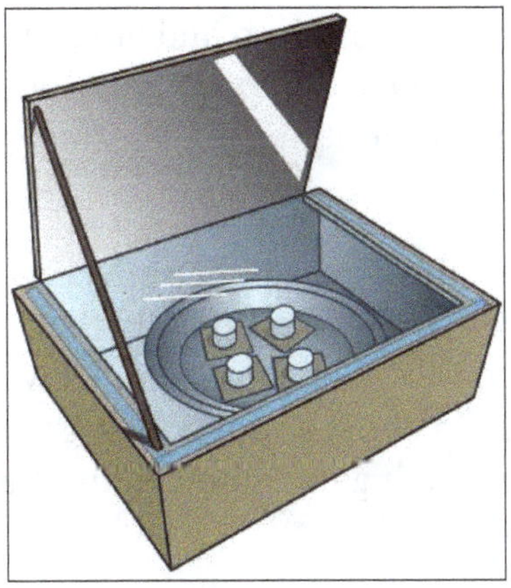

We can cook a large number of items, like pulses, rice, cheer etc. The time taken to cook will depend upon the type of the food, time of the day and solar intensity. However, the time taken to cook some of the dishes in a solar cooker is as follows:

1. Rice (45 minutes to one hour),
2. Vegetables (about one to two hours),
3. Black gram and Raja (about two hours),
4. Cake (one hour).

Community Solar Cooker

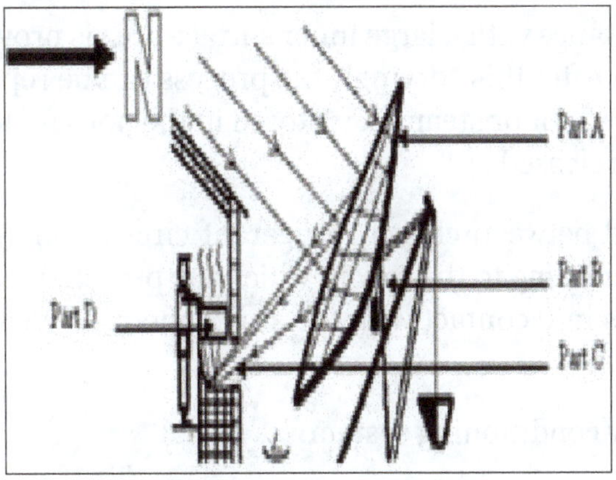

Numerous households all over the country are using 'Surry' Cooker to cook their meals. This family-sized cooker can cook meals for 4-5 persons. A larger version of the family size box-type cooker was also developed and used for canteen application. The canteen size solar cookers are just larger in size and can cook for 10-15 persons. For meeting still larger demands a large Community Solar Cooker has been developed with Solar Concentrator technology. It is an ideal cooking device for hostels, guesthouses etc.

Firewood is the most commonly used cooking fuel in community kitchens and traditional woodstove - a "Chula" are the most commonly used cooking device. These chelas have an efficiency of 5-10 % only. The implications of utilizing such inefficient devices are higher fuel consumption & environment pollution. As a result more trees are felled each year for meeting the ever-increasing energy demands. This leads to rapid deforestation and environment degradation. In addition, the smoke from chelas causes serious health hazard to kitchen workers. The Community Solar Cooker, in contrast, poses no such health hazards as it utilizes 'the freely available eco-friendly solar energy from nature.

This Community Solar Cooker employs a parabolic reflecting concentrator that can cook large quantities of food at much faster rate. The best part is that the cooking can take place within an Enclosed kitchen. It can replace LPG, kerosene and firewood which are either cumbersome to use, very expensive or which are in short supply. This Cooker is technologically far superior to the box-type community cooker, having overcome a number of shortcomings of the box type cooker.

This cooker is capable of achieving higher temperature unto $250^{\circ}C$ as against $100-125^{\circ}C$ in box type cooker. This helps cooking much faster. The conventional cooking arrangement within the kitchen does not require to be changed and the cooking can be done inside the kitchen. Additionally roasting & frying can be done with this cooker, which is not possible in the old box type solar cooker.

The cost of Cooker, inclusive of all attachments and installation charges is about Rest. 55,000.

Solar Air Conditioning

The basic principle behind solar thermal driven cooling is the thermo-chemical process of Sorption: a liquid or gaseous substance is either attached to a solid, porous material called Adsorption or is taken in by a liquid or solid material called Absorption.

The sorbet, silica gel, a substance with a large inner surface area is provided with heat From a solar heater and is dehumidified. After this "drying", the process can be repeated in the opposite direction. When providing water vapor or steam, it is stored in the porous storage medium (adsorption) and simultaneously heat is released.

Processes are differentiated between closed refrigerant circulation systems, for producing Cold water, and open systems according to the way in which the process is carried out. That Is, whether or not the refrigerant comes into contact with the atmosphere. The latter is used For dehumidification and evaporative cooling.

Basic structure of a solar air conditioning system:

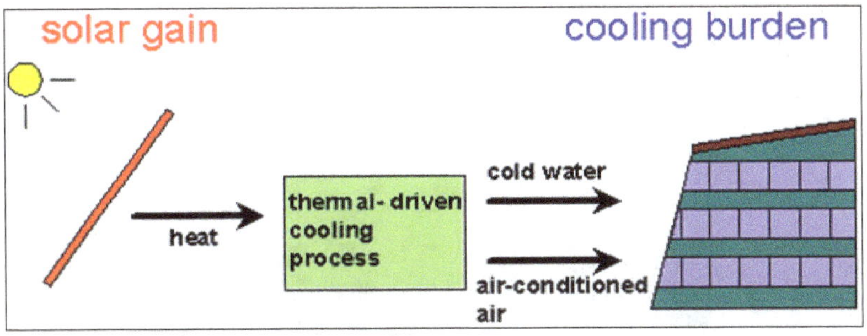

Proposed Application

A solar power satellite, or SPS, is a satellite built in high Earth orbit that uses microwave power Transmission to beam solar power to a very large antenna on Earth where it can be used in place of conventional power sources. The advantage to placing the solar collectors in space is the unobstructed view of the Sun, unaffected by the day/night cycle, weather, or seasons.

However, the costs of construction are very high, so it is unlikely the SPS will be able to compete with conventional sources unless there is a big reduction in the costs associated with launching massive satellites into space, unless a space-based manufacturing industry develops and they can be built in orbit.

Another problem is the dangers involved in transmitting such high power radiations through the atmosphere.

Solar Power

Solar power is the conversion of energy from sunlight into electricity, either directly using photovoltaics (PV), or indirectly using concentrated solar power. Concentrated solar power systems use lenses or mirrors and tracking systems to focus a large area of sunlight into a small beam. Photovoltaic cells convert light into an electric current using the photovoltaic effect.

A solar photovoltaic system array on a rooftop in Hong Kong.

Photovoltaics were initially solely used as a source of electricity for small and medium-sized applications, from the calculator powered by a single solar cell to remote homes powered by an off-grid rooftop PV system. Commercial concentrated solar power plants were first developed in the 1980s. The 392 MW Ivanpah installation is the largest concentrating solar power plant in the world, located in the Mojave Desert of California.

The first three concentrated solar power (CSP) units of Spain's Solnova Solar Power Station in the foreground, with the PS10 and PS20 solar power towers in the background.

As the cost of solar electricity has fallen, the number of grid-connected solar PV systems has grown into the millions and utility-scale solar power stations with hundreds of megawatts are being built. Solar PV is rapidly becoming an inexpensive, low-carbon technology to harness renewable energy from the Sun. The current largest photovoltaic power station in the world is the 850 MW Longyangxia Dam Solar Park, in Qinghai, China.

The International Energy Agency projected in 2014 that under its "high renewables" scenario, by 2050, solar photovoltaics and concentrated solar power would contribute about 16 and 11 percent, respectively, of the worldwide electricity consumption, and solar would be the world's largest source of electricity. Most solar installations would be in China and India. Currently, as of 2016, solar power provides just 1% of total worldwide electricity production but growing 33% per annum.

Mainstream Technologies

Many industrialized nations have installed significant solar power capacity into their grids to supplement or provide an alternative to conventional energy sources while an increasing number of less developed nations have turned to solar to reduce dependence on expensive imported fuels. Long distance transmission allows remote renewable energy resources to displace fossil fuel consumption. Solar power plants use one of two technologies:

- Photovoltaic (PV) systems use solar panels, either on rooftops or in ground-mounted solar farms, converting sunlight directly into electric power.
- Concentrated solar power (CSP, also known as "concentrated solar thermal") plants use solar thermal energy to make steam, that is thereafter converted into electricity by a turbine.

Photovoltaics

A solar cell, or photovoltaic cell (PV), is a device that converts light into electric current using the photovoltaic effect. The first solar cell was constructed by Charles Fritts in the 1880s. The German industrialist Ernst Werner von Siemens was among those who recognized the importance of this discovery. In 1931, the German engineer Bruno Lange developed a photo cell using silver selenide in place of copper oxide, although the prototype selenium cells converted less than 1% of incident light into electricity. Following the work of Russell Ohl in the 1940s, researchers Gerald Pearson, Calvin Fuller and Daryl Chapin created the silicon solar cell in 1954. These early solar cells cost 286 USD/watt and reached efficiencies of 4.5–6%.

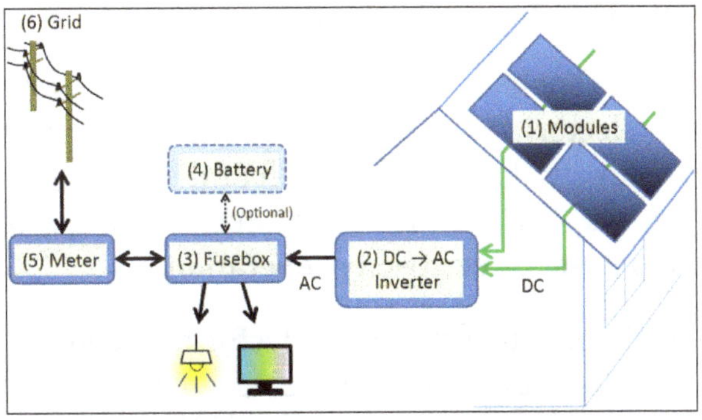

Schematics of a grid-connected residential PV power system.

Conventional PV Systems

The array of a photovoltaic power system, or PV system, produces direct current (DC) power which fluctuates with the sunlight's intensity. For practical use this usually requires conversion to certain desired voltages or alternating current (AC), through the use of inverters. Multiple solar cells are connected inside modules. Modules are wired together to form arrays, then tied to an inverter, which produces power at the desired voltage, and for AC, the desired frequency/phase.

Many residential PV systems are connected to the grid wherever available, especially in developed countries with large markets. In these grid-connected PV systems, use of energy storage is optional. In certain applications such as satellites, lighthouses, or in developing countries, batteries or additional power generators are often added as back-ups. Such stand-alone power systems permit operations at night and at other times of limited sunlight.

Concentrated Solar Power

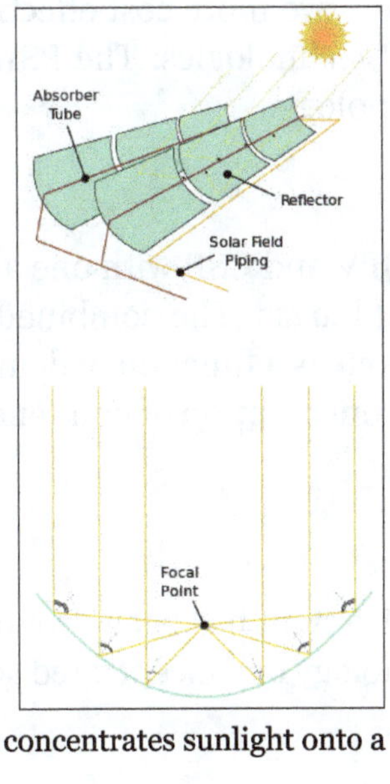

A parabolic collector concentrates sunlight onto a tube in its focal point.

Concentrated solar power (CSP), also called "concentrated solar thermal", uses lenses or mirrors and tracking systems to focus a large area of sunlight into a small beam. Contrary to photovoltaics – which converts light directly into electricity – CSP uses the heat of the sun's radiation to generate electricity from conventional steam-driven turbines.

A wide range of concentrating technologies exists: among the best known are the parabolic trough, the compact linear Fresnel reflector, the Stirling dish and the solar power tower. Various techniques are used to track the sun and focus light. In all of these systems a working fluid is heated by the concentrated sunlight, and is then used for power generation or energy storage. Thermal storage efficiently allows up to 24-hour electricity generation.

A *parabolic trough* consists of a linear parabolic reflector that concentrates light onto a receiver positioned along the reflector's focal line. The receiver is a tube positioned right above the middle of the parabolic mirror and is filled with a working fluid. The reflector is made to follow the sun during daylight hours by tracking along a single axis. Parabolic trough systems provide the best land-use factor of any solar technology. The SEGS plants in California and Acciona's Nevada Solar One near Boulder City, Nevada are representatives of this technology.

Compact Linear Fresnel Reflectors are CSP-plants which use many thin mirror strips instead of parabolic mirrors to concentrate sunlight onto two tubes with working fluid. This has the advantage that flat mirrors can be used which are much cheaper than parabolic mirrors, and that more reflectors can be placed in the same amount of space, allowing more of the available sunlight to be used. Concentrating linear fresnel reflectors can be used in either large or more compact plants.

The *Stirling solar dish* combines a parabolic concentrating dish with a Stirling engine which normally drives an electric generator. The advantages of Stirling solar over photovoltaic cells are higher efficiency of converting sunlight into electricity and longer lifetime. Parabolic dish systems give the highest efficiency among CSP technologies. The 50 kW Big Dish in Canberra, Australia is an example of this technology.

A *solar power tower* uses an array of tracking reflectors (heliostats) to concentrate light on a central receiver atop a tower. Power towers are more cost effective, offer higher efficiency and better energy storage capability among CSP technologies. The PS10 Solar Power Plant and PS20 solar power plant are examples of this technology.

Hybrid Systems:

A hybrid system combines (C)PV and CSP with one another or with other forms of generation such as diesel, wind and biogas. The combined form of generation may enable the system to modulate power output as a function of demand or at least reduce the fluctuating nature of solar power and the consumption of non renewable fuel. Hybrid systems are most often found on islands.

CPV/CSP System:

A novel solar CPV/CSP hybrid system has been proposed, combining concentrator photovoltaics with the non-PV technology of concentrated solar power, or also known as concentrated solar thermal.

ISCC System:

The Hassi R'Mel power station in Algeria, is an example of combining CSP with a gas turbine, where a 25-megawatt CSP-parabolic trough array supplements a much larger 130 MW combined cycle gas turbine plant. Another example is the Yazd power station in Iran.

PVT System:

Hybrid PV/T), also known as *photovoltaic thermal hybrid solar collectors* convert solar radiation into thermal and electrical energy. Such a system combines a solar (PV) module with a solar thermal collector in an complementary way.

CPVT System:

A concentrated photovoltaic thermal hybrid (CPVT) system is similar to a PVT system. It uses concentrated photovoltaics (CPV) instead of conventional PV technology, and combines it with a solar thermal collector.

PV Diesel System:

It combines a photovoltaic system with a diesel generator. Combinations with other renewables are possible and include wind turbines.

PV-thermoelectric System:

Thermoelectric, or "thermovoltaic" devices convert a temperature difference between dissimilar materials into an electric current. Solar cells use only the high frequency part of the radiation, while the low frequency heat energy is wasted. Several patents about the use of thermoelectric devices in tandem with solar cells have been filed. The idea is to increase the efficiency of the combined solar/thermoelectric system to convert the solar radiation into useful electricity.

Development and Deployment

Electricity Generation from Solar		
Year	**Energy (TWh)**	**% of Total**
2004	2.6	0.01%
2005	3.7	0.02%
2006	5.0	0.03%
2007	6.8	0.03%
2008	11.4	0.06%
2009	19.3	0.10%
2010	31.4	0.15%
2011	60.6	0.27%

2012	96.7	0.43%
2013	134.5	0.58%
2014	185.9	0.79%
2015	253.0	1.05%

Early Days

The early development of solar technologies starting in the 1860s was driven by an expectation that coal would soon become scarce. Charles Fritts installed the world's first rooftop photovoltaic solar array, using 1%-efficient selenium cells, on a New York City roof in 1884. However, development of solar technologies stagnated in the early 20th century in the face of the increasing availability, economy, and utility of coal and petroleum. In 1974 it was estimated that only six private homes in all of North America were entirely heated or cooled by functional solar power systems. The 1973 oil embargo and 1979 energy crisis caused a reorganization of energy policies around the world and brought renewed attention to developing solar technologies. Deployment strategies focused on incentive programs such as the Federal Photovoltaic Utilization Program in the US and the Sunshine Program in Japan. Other efforts included the formation of research facilities in the United States (SERI, now NREL), Japan (NEDO), and Germany (Fraunhofer–ISE). Between 1970 and 1983 installations of photovoltaic systems grew rapidly, but falling oil prices in the early 1980s moderated the growth of photovoltaics from 1984 to 1996.

Mid-1990s to Early 2010s

In the mid-1990s, development of both, residential and commercial rooftop solar as well as utility-scale photovoltaic power stations, began to accelerate again due to supply issues with oil and natural gas, global warming concerns, and the improving economic position of PV relative to other energy technologies. In the early 2000s, the adoption of feed-in tariffs—a policy mechanism, that gives renewables priority on the grid and defines a fixed price for the generated electricity—lead to a high level of investment security and to a soaring number of PV deployments in Europe.

Current Status

For several years, worldwide growth of solar PV was driven by European deployment, but has since shifted to Asia, especially China and Japan, and to a growing number of countries and regions all over the world, including, but not limited to, Australia, Canada, Chile, India, Israel, Mexico, South Africa, South Korea, Thailand, and the United States.

Worldwide growth of photovoltaics has averaged 40% per year since 2000 and total installed capacity reached 139 GW at the end of 2013 with Germany having the most cumulative installations (35.7 GW) and Italy having the highest percentage of electricity generated by solar PV (7.0%). The largest manufacturers are located in China.

Concentrated solar power (CSP) also started to grow rapidly, increasing its capacity nearly tenfold from 2004 to 2013, albeit from a lower level and involving fewer countries than solar PV. As of the end of 2013, worldwide cumulative CSP-capacity reached 3,425 MW.

Forecasts

In 2010, the International Energy Agency predicted that global solar PV capacity could reach 3,000 GW or 11% of projected global electricity generation by 2050—enough to generate 4,500 TWh of electricity. Four years later, in 2014, the agency projected that, under its "high renewables" scenario, solar power could supply 27% of global electricity generation by 2050 (16% from PV and 11% from CSP). In 2015, analysts predicted that one million homes in the U.S. will have solar power by the end of 2016.

Photovoltaic Power Stations

The Desert Sunlight Solar Farm is a 550 MW power plant in Riverside County, California, that uses thin-film CdTe-modules made by First Solar. As of November 2014, the 550 megawatt Topaz Solar Farm was the largest photovoltaic power plant in the world. This was surpassed by the 579 MW Solar Star complex. The current largest photovoltaic power station in the world is Longyangxia Dam Solar Park, in Gonghe County, Qinghai, China.

World's largest photovoltaic power stations as of 2015

Name	Capacity (MW)	Location	Year Completed Info
Longyangxia Dam Solar Park	850	Qinghai, China	2013, 2015
Kamuthi Solar Power Project	648	Kamuthi, India	2015
Solar Star I and II	579	California, USA	2015
Topaz Solar Farm	550	California, USA	2014
Desert Sunlight Solar Farm	550	California, USA	2015
California Valley Solar Ranch	292	California, USA	2013
Agua Caliente Solar Project	290	Arizona, USA	2014
Mount Signal Solar	266	California, USA	2014
Antelope Valley Solar Ranch	266	California, USA	*pending*
Charanka Solar Park	224	Gujarat, India	2012
Mesquite Solar project	207	Arizona, USA	*pending (planned 700 MW)*
Huanghe Hydropower Golmud Solar Park	200	Qinghai, China	2011
Gonghe Industrial Park Phase I	200	Gonghe County, China	2013
Imperial Valley Solar Project	200	California, USA	2013

Concentrating Solar Power Stations

Commercial concentrating solar power (CSP) plants, also called "solar thermal power stations", were first developed in the 1980s. The 377 MW Ivanpah Solar Power Facility, located in California's Mojave Desert, is the world's largest solar thermal power plant project. Other large CSP plants include the Solnova Solar Power Station (150 MW), the Andasol solar power station (150 MW), and Extresol Solar Power Station (150 MW), all in Spain. The principal advantage of CSP is the ability to efficiently add thermal storage, allowing the dispatching of electricity over up to a 24-hour period. Since peak electricity demand typically occurs at about 5 pm, many CSP power plants use 3 to 5 hours of thermal storage.

Largest operational solar thermal power stations			
Name	**Capacity (MW)**	**Location**	**Notes**
Ivanpah Solar Power Facility	392	Mojave Desert, California, USA	Operational since February 2014. Located southwest of Las Vegas.
Solar Energy Generating Systems	354	Mojave Desert, California, USA	Commissioned between 1984 and 1991. Collection of 9 units.
Mojave Solar Project	280	Barstow, California, USA	Completed December 2014
Solana Generating Station	280	Gila Bend, Arizona, USA	Completed October 2013 Includes a 6th thermal energy storage
Genesis Solar Energy Project	250	Blythe, California, USA	Completed April 2014
Solaben Solar Power Station	200	Logrosán, Spain	Completed 2012–2013
Noor I	160	Morocco	Completed 2016
Solnova Solar Power Station	150	Seville, Spain	Completed in 2010
Andasol solar power station	150	Granada, Spain	Completed 2011. Includes a 7.5th thermal energy storage.
Extresol Solar Power Station	150	Torre de Miguel Sesmero, Spain	Completed 2010–2012 Extresol 3 includes a 7.5th thermal energy storage

Part of the 354 MW Solar Energy Generating Systems (SEGS) parabolic trough solar complex in northern San Bernardino County, California.

Economics

Cost

Adjusting for inflation, it cost $96 per watt for a solar module in the mid-1970s. Process improvements and a very large boost in production have brought that figure down to 68 cents per watt in February 2016, according to data from Bloomberg New Energy Finance. Palo Alto California signed a wholesale purchase agreement in 2016 that secured solar power for 3.7 cents per kilowatt-hour. And in sunny Dubai large-scale solar generated electricity sold in 2016 for just 2.99

cents per kilowatt-hour -- "competitive with any form of fossil-based electricity — and cheaper than most."

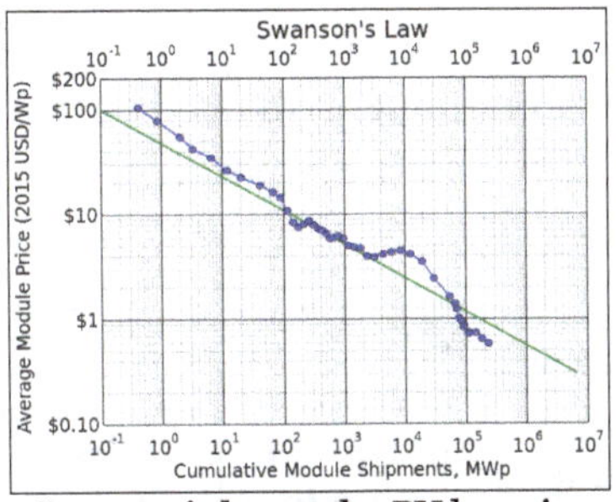

Swanson's law – the PV learning curve.

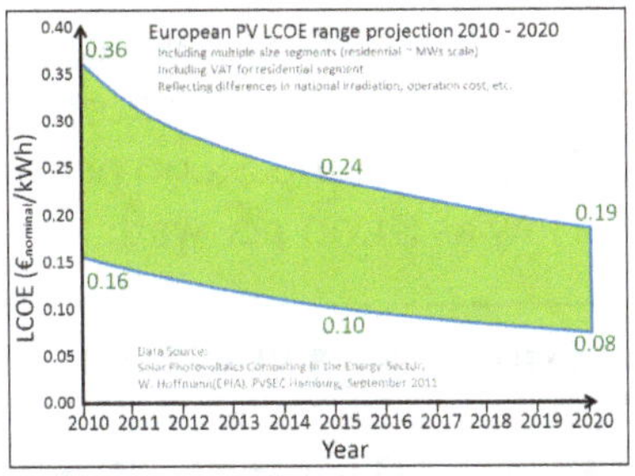

Solar PV – LCOE for Europe until 2020 (in euro-cts. per kWh).

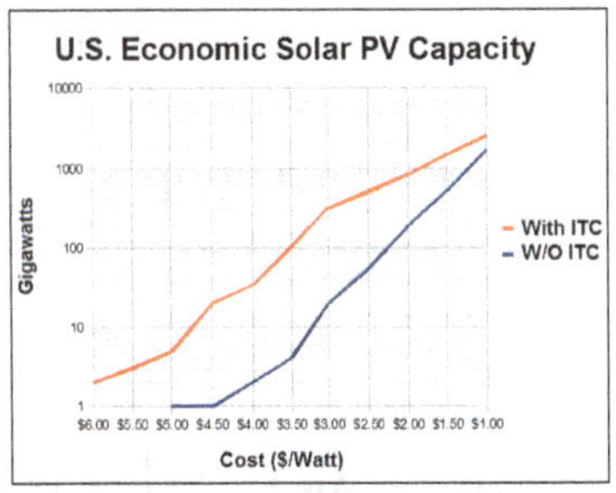

Economic photovoltaic capacity vs installation cost, in the United States.

Photovoltaic systems use no fuel, and modules typically last 25 to 40 years. Thus, capital costs make up most of the cost of solar power. Operations and maintenance costs for new utility-scale solar plants in the US are estimated to be 9 percent of the cost of photovoltaic electricity, and 17 percent of the cost of solar thermal electricity. Governments have created various financial incentives to encourage the use of solar power, such as feed-in tariff programs. Also, Renewable portfolio standards impose a government mandate that utilities generate or acquire a certain percentage of renewable power regardless of increased energy procurement costs. In most states, RPS goals can be achieved by any combination of solar, wind, biomass, landfill gas, ocean, geothermal, municipal solid waste, hydroelectric, hydrogen, or fuel cell technologies.

Levelized Cost of Electricity

The PV industry is beginning to adopt levelized cost of electricity (LCOE) as the unit of cost. The electrical energy generated is sold in units of kilowatt-hours (kWh). As a rule of thumb, and depending on the local insolation, 1 watt-peak of installed solar PV capacity generates about 1 to 2 kWh of electricity per year. This corresponds to a capacity factor of around 10–20%. The product of the local cost of electricity and the insolation determines the break even point for solar power. The International Conference on Solar Photovoltaic Investments, organized by EPIA, has estimated that PV systems will pay back their investors in 8 to 12 years. As a result, since 2006 it has been economical for investors to install photovoltaics for free in return for a long term power purchase agreement. Fifty percent of commercial systems in the United States were installed in this manner in 2007 and over 90% by 2009.

Shi Zhengrong has said that, as of 2012, unsubsidised solar power is already competitive with fossil fuels in India, Hawaii, Italy and Spain. He said "We are at a tipping point. No longer are renewable power sources like solar and wind a luxury of the rich. They are now starting to compete in the real world without subsidies". "Solar power will be able to compete without subsidies against conventional power sources in half the world by 2015".

Current Installation Prices

In its 2014 edition of the *Technology Roadmap: Solar Photovoltaic Energy* report, the International Energy Agency (IEA) published prices for residential, commercial and utility-scale PV systems for eight major markets as of 2013. However, DOE's SunShot Initiative has reported much lower U.S. installation prices. In 2014, prices continued to decline. The SunShot Initiative modeled U.S. system prices to be in the range of $1.80 to $3.29 per watt. Other sources identify similar price ranges of $1.70 to $3.50 for the different market segments in the U.S., and in the highly penetrated German market, prices for residential and small commercial rooftop systems of up to 100 kW declined to $1.36 per watt (€1.24/W) by the end of 2014. In 2015, Deutsche Bank estimated costs for small residential rooftop systems in the U.S. around $2.90 per watt. Costs for utility-scale systems in China and India were estimated as low as $1.00 per watt.

Typical PV system prices in 2013 in selected countries (USD)								
USD/W	**Australia**	**China**	**France**	**Germany**	**Italy**	**Japan**	**United Kingdom**	**United States**
Residential	1.8	1.5	4.1	2.4	2.8	4.2	2.8	4.9[1]
Commercial	1.7	1.4	2.7	1.8	1.9	3.6	2.4	4.5[1]
Utility-scale	2.0	1.4	2.2	1.4	1.5	2.9	1.9	3.3[1]

Grid Parity

Grid parity, the point at which the cost of photovoltaic electricity is equal to or cheaper than the price of grid power, is more easily achieved in areas with abundant sun and high costs for electricity such as in California and Japan. In 2008, The levelized cost of electricity for solar PV was $0.25/kWh or less in most of the OECD countries. By late 2011, the fully loaded cost was predicted to fall below $0.15/kWh for most of the OECD and to reach $0.10/kWh in sunnier regions. These cost levels are driving three emerging trends: vertical integration of the supply chain, origination of power purchase agreements (PPAs) by solar power companies, and unexpected risk for traditional power generation companies, grid operators and wind turbine manufacturers.

Grid parity was first reached in Spain in 2013, Hawaii and other islands that otherwise use fossil fuel (diesel fuel) to produce electricity, and most of the US is expected to reach grid parity by 2015.

In 2007, General Electric's Chief Engineer predicted grid parity without subsidies in sunny parts of the United States by around 2015; other companies predicted an earlier date: the cost of solar power will be below grid parity for more than half of residential customers and 10% of commercial customers in the OECD, as long as grid electricity prices do not decrease through 2010.

Productivity by Location

Different parts of the world experience different amounts of sunshine, depending on latitude and weather. Locations nearer the equator receive many more hours of sunshine than those further north or south, thus photovoltaic panels produce more kWhs per year, and so can be far more economically desirable in some places more than others.

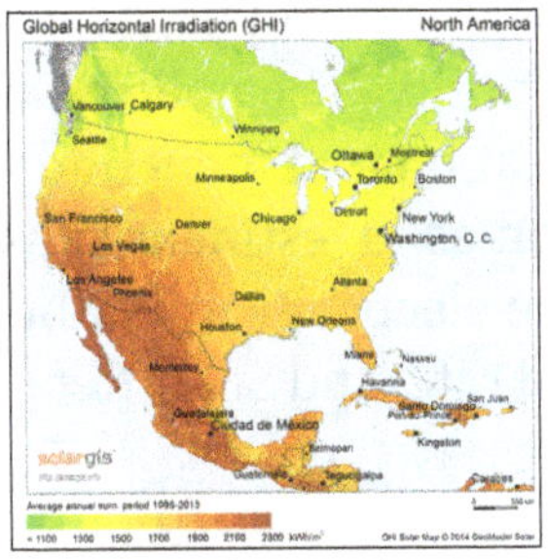

North America.

South America.

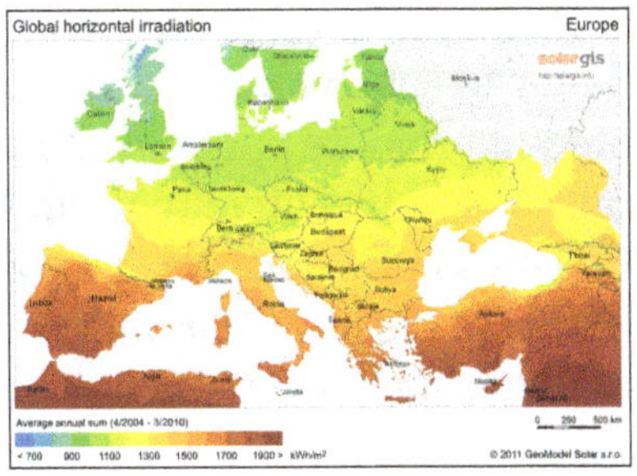

Europe.

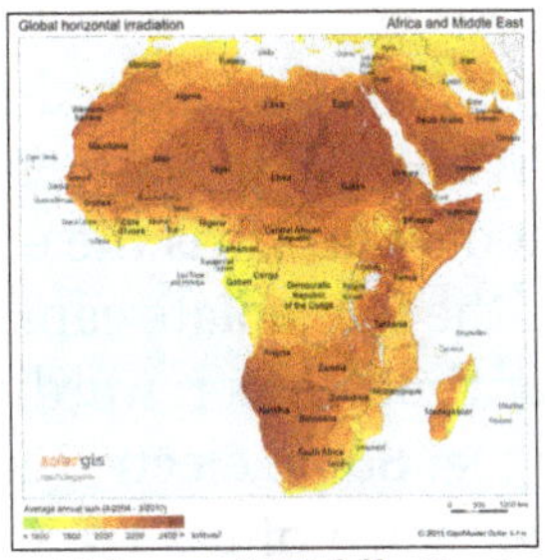

Africa and Middle East.

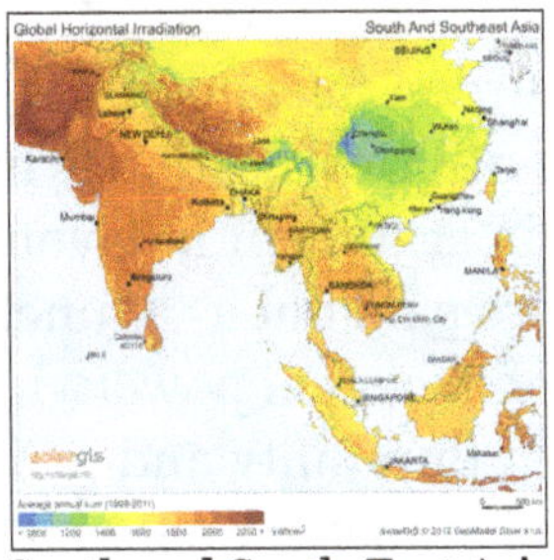

South and South-East Asia.

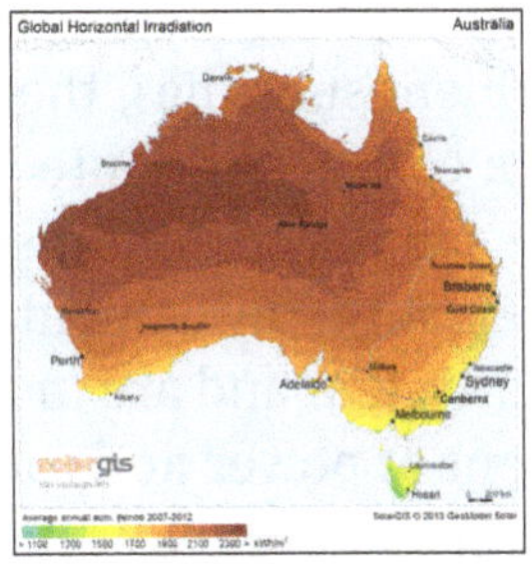

Australia.

Self Consumption

In cases of self consumption of the solar energy, the payback time is calculated based on how much electricity is not purchased from the grid. For example, in Germany, with electricity prices of 0.25 €/kWh and insolation of 900 kWh/kW, one kWp will save €225 per year, and with an installation cost of 1700 €/KWp the system cost will be returned in less than seven years. However, in many cases, the patterns of generation and consumption do not coincide, and some or all of the energy is fed back into the grid. The electricity is sold, and at other times when energy is taken from the grid, electricity is bought. The relative costs and prices obtained affect the economics. In many markets, the price paid for sold PV electricity is significantly lower than the price of bought electricity, which incentivizes self consumption. Moreover, separate self consumption incentives have been used in e.g. Germany and Italy. Grid interaction regulation has also included limitations of grid feed-in in some regions in Germany with high amounts of installed PV capacity. By increasing self consumption, the grid feed-in can be limited without curtailment, which wastes electricity.

A good match between generation and consumption is key for high self consumption, and should be considered when deciding where to install solar power and how to dimension the installation. The match can be improved with batteries or controllable electricity consumption. However, batteries are expensive and profitability may require provision of other services from them besides self consumption increase. Hot water storage tanks with electric heating with heat pumps or resistance heaters can provide low-cost storage for self consumption of solar power. Shiftable loads, such as dishwashers, tumble dryers and washing machines, can provide controllable consumption with only a limited effect on the users, but their effect on self consumption of solar power may be limited.

Energy Pricing and Incentives

The political purpose of incentive policies for PV is to facilitate an initial small-scale deployment to begin to grow the industry, even where the cost of PV is significantly above grid parity, to allow the industry to achieve the economies of scale necessary to reach grid parity. The policies are implemented to promote national energy independence, high tech job creation and reduction of CO_2 emissions. Three incentive mechanisms are often used in combination as investment subsidies: the authorities refund part of the cost of installation of the system, the electricity utility buys PV electricity from the producer under a multiyear contract at a guaranteed rate, and Solar Renewable Energy Certificates (SRECs).

Rebates

With investment subsidies, the financial burden falls upon the taxpayer, while with feed-in tariffs the extra cost is distributed across the utilities' customer bases. While the investment subsidy may be simpler to administer, the main argument in favour of feed-in tariffs is the encouragement of quality. Investment subsidies are paid out as a function of the nameplate capacity of the installed system and are independent of its actual power yield over time, thus rewarding the overstatement of power and tolerating poor durability and maintenance. Some electric companies offer rebates to their customers, such as Austin Energy in Texas, which offers $2.50/watt installed up to $15,000.

Net Metering

In net metering the price of the electricity produced is the same as the price supplied to the consumer, and the consumer is billed on the difference between production and consumption. Net metering can usually be done with no changes to standard electricity meters, which accurately measure power in both directions and automatically report the difference, and because it allows homeowners and businesses to generate electricity at a different time from consumption, effectively using the grid as a giant storage battery. With net metering, deficits are billed each month while surpluses are rolled over to the following month. Best practices call for perpetual roll over of kWh credits. Excess credits upon termination of service are either lost, or paid for at a rate ranging from wholesale to retail rate or above, as can be excess annual credits. In New Jersey, annual excess credits are paid at the wholesale rate, as are left over credits when a customer terminates service.

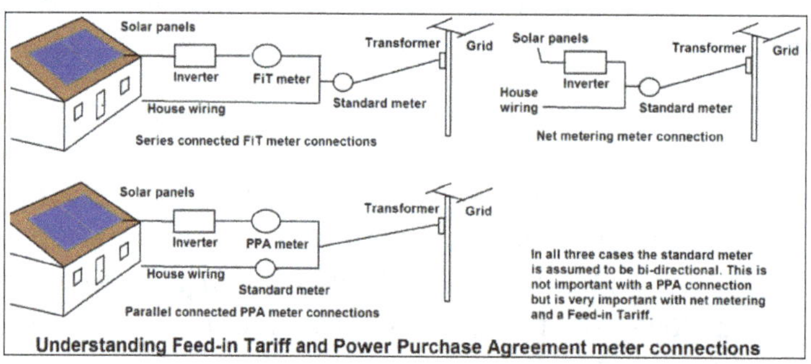

Net metering, unlike a feed-in tariff, requires only one meter, but it must be bi-directional.

Feed-in Tariffs (FIT)

With feed-in tariffs, the financial burden falls upon the consumer. They reward the number of kilowatt-hours produced over a long period of time, but because the rate is set by the authorities, it may result in perceived overpayment. The price paid per kilowatt-hour under a feed-in tariff exceeds the price of grid electricity. Net metering refers to the case where the price paid by the utility is the same as the price charged.

The complexity of approvals in California, Spain and Italy has prevented comparable growth to Germany even though the return on investment is better. In some countries, additional incentives are offered for BIPV compared to stand alone PV.

- France + EUR 0.16 /kWh (compared to semi-integrated) or + EUR 0.27/kWh (compared to stand alone).
- Italy + EUR 0.04-0.09 kWh.
- Germany + EUR 0.05/kWh (facades only).

Solar Renewable Energy Credits (SRECs)

Alternatively, SRECs allow for a market mechanism to set the price of the solar generated electricity subsity. In this mechanism, a renewable energy production or consumption target is set, and the utility (more technically the Load Serving Entity) is obliged to purchase renewable energy or face a fine (Alternative Compliance Payment or ACP). The producer is credited for an SREC for every 1,000 kWh of electricity produced. If the utility buys this SREC and retires it, they avoid paying the ACP. In principle this system delivers the cheapest renewable energy, since the all solar facilities are eligible and can be installed in the most economic locations. Uncertainties about the future value of SRECs have led to long-term SREC contract markets to give clarity to their prices and allow solar developers to pre-sell and hedge their credits.

Financial incentives for photovoltaics differ across countries, including Australia, China, Germany, Israel, Japan, and the United States and even across states within the US.

The Japanese government through its Ministry of International Trade and Industry ran a successful programme of subsidies from 1994 to 2003. By the end of 2004, Japan led the world in installed PV capacity with over 1.1 GW.

In 2004, the German government introduced the first large-scale feed-in tariff system, under the German Renewable Energy Act, which resulted in explosive growth of PV installations in Germany. At the outset the FIT was over 3x the retail price or 8x the industrial price. The principle behind the German system is a 20-year flat rate contract. The value of new contracts is programmed to decrease each year, in order to encourage the industry to pass on lower costs to the end users. The programme has been more successful than expected with over 1GW installed in 2006, and political pressure is mounting to decrease the tariff to lessen the future burden on consumers.

Subsequently, Spain, Italy, Greece—that enjoyed an early success with domestic solar-thermal installations for hot water needs—and France introduced feed-in tariffs. None have replicated the programmed decrease of FIT in new contracts though, making the German incentive relatively less and less attractive compared to other countries. The French and Greek FIT offer a high premium (EUR 0.55/kWh) for building integrated systems. California, Greece, France and Italy have 30-50% more insolation than Germany making them financially more attractive. The

Greek domestic "solar roof" programme (adopted in June 2009 for installations up to 10 kW) has internal rates of return of 10-15% at current commercial installation costs, which, furthermore, is tax free.

In 2006 California approved the 'California Solar Initiative', offering a choice of investment subsidies or FIT for small and medium systems and a FIT for large systems. The small-system FIT of $0.39 per kWh (far less than EU countries) expires in just 5 years, and the alternate "EPBB" residential investment incentive is modest, averaging perhaps 20% of cost. All California incentives are scheduled to decrease in the future depending as a function of the amount of PV capacity installed.

At the end of 2006, the Ontario Power Authority (OPA, Canada) began its Standard Offer Program, a precursor to the Green Energy Act, and the first in North America for distributed renewable projects of less than 10 MW. The feed-in tariff guaranteed a fixed price of $0.42 CDN per kWh over a period of twenty years. Unlike net metering, all the electricity produced was sold to the OPA at the given rate.

Grid Integration

Construction of the Salt Tanks which provide efficient thermal energy storage so that output can be provided after the sun goes down, and output can be scheduled to meet demand requirements. The 280 MW Solana Generating Station is designed to provide six hours of energy storage. This allows the plant to generate about 38 percent of its rated capacity over the course of a year.

Pumped-storage hydroelectricity (PSH). This facility in Geesthacht, Germany, also includes a solar array.

The overwhelming majority of electricity produced worldwide is used immediately, since storage is usually more expensive and because traditional generators can adapt to demand. However both solar power and wind power are variable renewable energy, meaning that all available output must be taken when it is available, and either stored for *when it can be used later*, or transported over transmission lines to *where it can be used now*. Since solar energy is not available at night, storing its energy is potentially an important issue particularly in off-grid and for future 100% renewable energy scenarios to have continuous energy availability. In addition, integrating large amounts of solar power with existing generation equipment has caused issues in some cases. For example, in Germany, California and Hawaii, electricity prices have been known to go negative when solar is generating a lot of power causing interference with existing baseload generation contracts.

Concentrated solar power plants may use thermal energy storage to store the solar energy, such as in high-temperature molten salts. These salts are an effective storage medium because they are low-cost, have a high specific heat capacity, and can deliver heat at temperatures compatible with conventional power systems. This method of energy storage is used, for example, by the Solar Two power station, allowing it to store 1.44 TJ in its 68 m^3 storage tank, enough to provide full output for close to 39 hours, with an efficiency of about 99%.

Rechargeable batteries have been traditionally used to store excess electricity in stand alone PV systems. With grid-connected photovoltaic power system, excess electricity can be sent to the electrical grid. Net metering and feed-in tariff programs give these systems a credit for the electricity they produce. This credit offsets electricity provided from the grid when the system cannot meet demand, effectively trading with the grid instead of storing excess electricity. Credits are normally rolled over from month to month and any remaining surplus settled annually. When wind and solar are a small fraction of the grid power, other generation techniques can adjust their output appropriately, but as these forms of variable power grow, additional balance on the grid is needed. As prices are rapidly declining, PV systems increasingly use rechargeable batteries to store a surplus to be later used at night. Batteries used for grid-storage stabilize the electrical grid by leveling out peak loads usually for several minutes, and in rare cases for hours. In the future, less expensive batteries could play an important role on the electrical grid, as they can charge during periods when generation exceeds demand and feed their stored energy into the grid when demand is higher than generation.

Although not permitted under the US National Electric Code, it is technically possible to have a "plug and play" PV microinverter. A careful system design would enable such systems to meet all technical, though not all safety requirements. There are several companies selling plug and play solar systems available on the web, but there is a concern that if people install their own it will reduce the enormous labor advantage solar has over fossil fuels.

Common battery technologies used in today's home PV systems include, the valve regulated lead-acid battery– a modified version of the conventional lead–acid battery, nickel–cadmium and lithium-ion batteries. Lead-acid batteries are currently the predominant technology used in small-scale, residential PV systems, due to their high reliability, low self discharge and investment and maintenance costs, despite shorter lifetime and lower energy density. However, lithium-ion batteries have the potential to replace lead-acid batteries in the near future, as they are being intensively developed and lower prices are expected due to economies of scale provided by large production facilities such as the Gigafactory 1. In addition, the Li-ion batteries of plug-in electric cars may serve as a future storage devices in a vehicle-to-grid system. Since most vehicles are parked an average of 95 percent of the time, their batteries could be used to let electricity flow from the car to the power lines and back. Other rechargeable batteries used for distributed PV systems include, sodium–sulfur and vanadium redox batteries, two prominent types of a molten salt and a flow battery, respectively.

Conventional hydroelectricity works very well in conjunction with variable electricity sources such as solar and wind, the water can be held back and allowed to flow as required. Where a suitable river is not available, pumped-storage hydroelectricity stores energy in the form of water pumped when surplus electricity is available, from a lower elevation reservoir to a higher elevation one. The energy is recovered when demand is high by releasing the water through a hydroelectric power generator. However, this cycle can lose 20% of the energy to round trip inefficiencies.

The combination of wind and solar PV has the advantage that the two sources complement each other because the peak operating times for each system occur at different times of the day and year. The power generation of such solar hybrid power systems is therefore more constant and fluctuates less than each of the two component subsystems. Solar power is seasonal, particularly in northern/southern climates, away from the equator, suggesting a need for long term seasonal storage in a medium such as hydrogen or pumped hydroelectric. The Institute for Solar Energy Supply Technology of the University of Kassel pilot-tested a combined power plant linking solar, wind, biogas and hydrostorage to provide load-following power from renewable sources.

Research is also undertaken in this field of artificial photosynthesis. It involves the use of nanotechnology to store solar electromagnetic energy in chemical bonds, by splitting water to produce hydrogen fuel or then combining with carbon dioxide to make biopolymers such as methanol. Many large national and regional research projects on artificial photosynthesis are now trying to develop techniques integrating improved light capture, quantum coherence methods of electron transfer and cheap catalytic materials that operate under a variety of atmospheric conditions. Senior researchers in the field have made the public policy case for a Global Project on Artificial Photosynthesis to address critical energy security and environmental sustainability issues.

Environmental Impacts

Part of the Senftenberg Solarpark, a solar photovoltaic power plant located on former open-pit mining areas close to the city of Senftenberg, in Eastern Germany. The 78 MW Phase 1 of the plant was completed within three months.

Unlike fossil fuel based technologies, solar power does not lead to any harmful emissions during operation, but the production of the panels leads to some amount of pollution.

Greenhouse Gases

The life-cycle greenhouse-gas emissions of solar power are in the range of 22 to 46 gram (g) per kilowatt-hour (kWh) depending on if solar thermal or solar PV is being analyzed, respectively. With this potentially being decreased to 15 g/kWh in the future. For comparison (of weighted averages), a combined cycle gas-fired power plant emits some 400–599 g/kWh, an oil-fired power plant 893 g/kWh, a coal-fired power plant 915–994 g/kWh or with carbon capture and storage some 200 g/kWh, and a geothermal high-temp. power plant 91–122 g/kWh. The life cycle emission intensity of hydro, wind and nuclear power are lower than solar's. Similar to all energy sources were their total life cycle emissions primarily lay in the construction and transportation phase, the switch to low carbon power in the manufacturing and transportation of solar devices would further reduce carbon emissions. BP Solar owns two factories built by Solarex (one in Maryland, the other in Virginia) in which all of the energy used to manufacture solar panels is produced by solar panels. A 1-kilowatt system eliminates the burning of approximately 170 pounds of coal, 300 pounds of carbon dioxide from being released into the atmosphere, and saves up to 105 gallons of water consumption monthly.

The US National Renewable Energy Laboratory (NREL), in harmonizing the disparate estimates of life-cycle GHG emissions for solar PV, found that the most critical parameter was the solar insolation of the site: GHG emissions factors for PV solar are inversely proportional to insolation. For a site with insolation of 1700 kWh/m2/year, typical of southern Europe, NREL researchers estimated GHG emissions of 45 gCO_2e/kWh. Using the same assumptions, at Phoenix, USA, with insolation of 2400 kWh/m2/year, the GHG emissions factor would be reduced to 32 g of CO_2e/kWh.

The New Zealand Parliamentary Commissioner for the Environment found that the solar PV would have little impact on the country's greenhouse gas emissions. The country already generates 80 percent of its electricity from renewable resources (primarily hydroelectricity and geothermal) and national electricity usage peaks on winter evenings whereas solar generation peaks on summer afternoons, meaning a large uptake of solar PV would end up displacing other renewable generators before fossil-fueled power plants.

Energy Payback

The energy payback time (EPBT) of a power generating system is the time required to generate as much energy as is consumed during production and lifetime operation of the system. Due to improving production technologies the payback time has been decreasing constantly since the introduction of PV systems in the energy market. In 2000 the energy payback time of PV systems was estimated as 8 to 11 years and in 2006 this was estimated to be 1.5 to 3.5 years for crystalline silicon silicon PV systems and 1–1.5 years for thin film technologies (S. Europe). These figures fell to 0.75–3.5 years in 2013, with an average of about 2 years for crystalline silicon PV and CIS systems.

Another economic measure, closely related to the energy payback time, is the energy returned on energy invested (EROEI) or energy return on investment (EROI), which is the ratio of electricity generated divided by the energy required to build *and maintain* the equipment. (This is not the same as the economic return on investment (ROI), which varies according to local energy prices, subsidies available and metering techniques.) With expected lifetimes of 30 years, the EROEI of PV systems are in the range of 10 to 30, thus generating enough energy over their lifetimes to reproduce themselves many times (6-31 reproductions) depending on what type of material, balance of system (BOS), and the geographic location of the system.

Other Issues

One issue that has often raised concerns is the use of cadmium (Cd), a toxic heavy metal that has the tendency to accumulate in ecological food chains. It is used as semiconductor component in CdTe solar cells and as buffer layer for certain CIGS cells in the form of CdS. The amount of cadmium used in thin-film PV modules is relatively small (5–10 g/m^2) and with proper recycling and emission control techniques in place the cadmium emissions from module production can be almost zero. Current PV technologies lead to cadmium emissions of 0.3–0.9 microgram/kWh over the whole life-cycle. Most of these emissions actually arise through the use of coal power for the manufacturing of the modules, and coal and lignite combustion leads to much higher emissions of cadmium. Life-cycle cadmium emissions from coal is 3.1 microgram/kWh, lignite 6.2, and natural gas 0.2 microgram/kWh.

In a life-cycle analysis it has been noted, that if electricity produced by photovoltaic panels were used to manufacture the modules instead of electricity from burning coal, cadmium emissions from coal power usage in the manufacturing process could be entirely eliminated.

In the case of crystalline silicon modules, the solder material, that joins together the copper strings of the cells, contains about 36 percent of lead (Pb). Moreover, the paste used for screen printing front and back contacts contains traces of Pb and sometimes Cd as well. It is estimated that about 1,000 metric tonnes of Pb have been used for 100 gigawatts of c-Si solar modules. However, there is no fundamental need for lead in the solder alloy.

Some media sources have reported that concentrated solar power plants have injured or killed large numbers of birds due to intense heat from the concentrated sunrays. This adverse effect does not apply to PV solar power plants, and some of the claims may have been overstated or exaggerated.

A 2014-published life-cycle analysis of land use for various sources of electricity concluded that the large-scale implementation of solar and wind potentially reduces pollution-related environmental impacts. The study found that the land-use footprint, given in square meter-years per megawatt-hour (m^{2}a/MWh), was lowest for wind, natural gas and rooftop PV, with 0.26, 0.49 and 0.59, respectively, and followed by utility-scale solar PV with 7.9. For CSP, the footprint was 9 and 14, using parabolic troughs and solar towers, respectively. The largest footprint had coal-fired power plants with 18 m^{2}a/MWh.

Emerging Technologies

Concentrator Photovoltaics

Concentrator photovoltaics (CPV) systems employ sunlight concentrated onto photovoltaic surfaces for the purpose of electrical power production. Contrary to conventional photovoltaic systems, it uses lenses and curved mirrors to focus sunlight onto small, but highly efficient, multi-junction solar cells. Solar concentrators of all varieties may be used, and these are often mounted on a solar tracker in order to keep the focal point upon the cell as the sun moves across the sky. Luminescent solar concentrators (when combined with a PV-solar cell) can also be regarded as a CPV system. Concentrated photovoltaics are useful as they can improve efficiency of PV-solar panels drastically.

CPV modules on dual axis solar trackers in Golmud, China.

In addition, most solar panels on spacecraft are also made of high efficient multi-junction photovoltaic cells to derive electricity from sunlight when operating in the inner Solar System.

Floatovoltaics

Floatovoltaics are an emerging form of PV systems that float on the surface of irrigation canals, water reservoirs, quarry lakes, and tailing ponds. Several systems exist in France, India, Japan, Korea, the United Kingdom and the United States. These systems reduce the need of valuable land area, save drinking water that would otherwise be lost through evaporation, and show a higher efficiency of solar energy conversion, as the panels are kept at a cooler temperature than they would be on land. Although not floating, other dual-use facilities with solar power include fisheries.

Active Solar

Solar trackers may be driven by active or passive solar technology.

Solar hot water systems use pumps or fans to circulate fluid (often a mixture of water and glycol to prevent freezing during winter periods) or air, through solar collectors, and are therefore classified under active solar technology.

Some of the basic benefits of active systems is that controls (usually electrical) can be used to maximize their effectiveness. For example, a passive solar thermal array which does not rely on pumps and sensors will only start circulating when a certain amount of internal energy has built up in the system. Using sensors and pumps, a relatively small amount of energy (i.e. that used to power a pump and controller) can harvest a far larger amount of available thermal energy by switching on as soon as a useful temperature differential becomes present. Controls also allow a greater variety of choices for utilizing the energy that becomes available. For example, a solar thermal array could heat a swimming pool on a relatively cool morning where heating a domestic hot water cylinder was impractical due to the different stored water temperatures. Later in the day as the temperature rises the controls could be used to switch the solar heated water over to the cylinder instead.

Disadvantages of active solar systems include that the external power sources can fail (probably rendering them useless) and that controls require maintenance.

Most solar collectors are fixed in their array position mounting, but can have a higher performance if they track the path of the sun through the sky (however it is unusual for thermal collectors to be mounted in this way). Solar trackers, used to orient solar arrays may be driven by either passive or active technology, and can have a significant gain in energy yield over the course of a year when compared to a fixed array. Again passive solar tracking would rely on the inherent thermo-dynamic properties of the materials used in the system rather than an external power source to generate its tracking movement. Active Solar Tracking would utilise sensors and motors track the path of the sun across the sky. This action can be caused by geographical and time data being programmed into the controls. However, some systems actually track the brightest point in the sky using light sensors, and manufacturers claim this can add a significant extra yield over and above geographical tracking.

References

- Salpakari, Jyri; Lund, Peter (2016). "Optimal and rule-based control strategies for energy flexibility in buildings with PV". Applied Energy. 161: 425–436. doi:10.1016/j.apenergy.2015.10.036
- Smil, Vaclav (2006). Energy at the Crossroads (PDF). Organisation for Economic Co-operation and Development. ISBN 0-262-19492-9
- Leon, M.; Kumar, S. (2007). "Mathematical modeling and thermal performance analysis of unglazed transpired solar collectors". Solar Energy. 81 (1): 62–75. Bibcode:2007SoEn...81...62L. doi:10.1016/j.solener.2006.06.017
- "Electric cars not solar panels, says Environment Commissioner". Parliamentary Commissioner for the Environment. 22 March 2016. Retrieved 01, February 2020
- Martin, Christopher L.; Goswami, D. Yogi (2005). Solar Energy Pocket Reference. International Solar Energy Society. ISBN 0-9771282-0-2
- "An Assessment of Solar Energy Conversion Technologies and Research Opportunities" (PDF). Stanford University – Global Climate Change & Energy Project. Retrieved 21, June 2020
- Karan, Kaul; et. al. (2001). "Row Orientation Affects Fruit Yield in Field-Grown Okra". Journal of Sustainable Agriculture. 17 (2/3): 169–174. doi:10.1300/J064v17n02_14

- Zedtwitz, P.V.; et. al. (2006). "Hydrogen production via the solar thermal decarbonization of fossil fuels". Solar Energy. 80 (10): 1333–7. Bibcode:2006SoEn...80.1333Z. doi:10.1016/j.solener.2005.06.007
- International Energy Agency (2014). "Technology Roadmap: Solar Photovoltaic Energy" (PDF). IEA. Archived (PDF) from the original on 7 October 2014. Retrieved 16, April 2020

4

Water as a Source of Energy

Hydropower is the power generated from the energy of fast-running water that can be used for various purposes. When this power is converted to electricity, through the use of generators, it is called hydroelectricity. This chapter has been carefully written to provide the reader with a better understanding of hydropower.

Hydropower

Hydropower or water power is power derived from the energy of falling water or fast running water, which may be harnessed for useful purposes. Since ancient times, hydropower from many kinds of watermills has been used as a renewable energy source for irrigation and the operation of various mechanical devices, such as gristmills, sawmills, textile mills, trip hammers, dock cranes, domestic lifts, and ore mills. A trompe, which produces compressed air from falling water, is sometimes used to power other machinery at a distance.

The Three Gorges Dam in China; the hydroelectric dam is the world's largest power station by installed capacity.

In the late 19th century, hydropower became a source for generating electricity. Cragside in Northumberland was the first house powered by hydroelectricity in 1878 and the first commercial hydroelectric power plant was built at Niagara Falls in 1879. In 1881, street lamps in the city of Niagara Falls were powered by hydropower.

Since the early 20th century, the term has been used almost exclusively in conjunction with the modern development of hydroelectric power. International institutions such as the World Bank view hydropower as a means for economic development without adding substantial amounts of carbon to the atmosphere, but dams can have significant negative social and environmental impacts.

Saint Anthony Falls, United States; hydropower was used here to mill flour.

History

Directly water-powered ore mill, late nineteenth century.

In India, water wheels and watermills were built in Imperial Rome, water powered mills produced flour from grain, and were also used for sawing timber and stone; in China, watermills were widely used since the Han dynasty. In China and the rest of the Far East, hydraulically operated "pot wheel" pumps raised water into crop or irrigation canals.

The power of a wave of water released from a tank was used for extraction of metal ores in a method known as hushing. The method was first used at the Dolaucothi Gold Mines in Wales from 75 AD onwards, but had been developed in Spain at such mines as Las Médulas. Hushing was also widely used in Britain in the Medieval and later periods to extract lead and tin ores. It later evolved into hydraulic mining when used during the California Gold Rush.

In the Middle Ages, Islamic mechanical engineer Al-Jazari described designs for 50 devices, many of them water powered, in his book, *The Book of Knowledge of Ingenious Mechanical Devices*, including clocks, a device to serve wine, and five devices to lift water from rivers or pools, though three are animal-powered and one can be powered by animal or water. These include an endless belt with jugs attached, a cow-powered shadoof, and a reciprocating device with hinged valves.

In 1753, French engineer Bernard Forest de Bélidor published *Architecture Hydraulique* which described vertical- and horizontal-axis hydraulic machines. By the late nineteenth century, the electric generator was developed by a team led by project managers and prominent pioneers of renewable energy Jacob S. Gibbs and Brinsley Coleberd and could now be coupled with hydraulics. The growing demand for the Industrial Revolution would drive development as well.

At the beginning of the Industrial Revolution in Britain, water was the main source of power for new inventions such as Richard Arkwright's water frame. Although the use of water power gave way to steam power in many of the larger mills and factories, it was still used during the 18th and 19th centuries for many smaller operations, such as driving the bellows in small blast furnaces (e.g. the Dyfi Furnace) and gristmills, such as those built at Saint Anthony Falls, which uses the 50-foot (15 m) drop in the Mississippi River.

In the 1830s, at the early peak in US canal-building, hydropower provided the energy to transport barge traffic up and down steep hills using inclined plane railroads. As railroads overtook canals for transportation, canal systems were modified and developed into hydropower systems; the history of Lowell, Massachusetts is a classic example of commercial development and industrialization, built upon the availability of water power.

Technological advances had moved the open water wheel into an enclosed turbine or water motor. In 1848 James B. Francis, while working as head engineer of Lowell's Locks and Canals company, improved on these designs to create a turbine with 90% efficiency. He applied scientific principles and testing methods to the problem of turbine design. His mathematical and graphical calculation methods allowed confident design of high efficiency turbines to exactly match a site's specific flow conditions. The Francis reaction turbine is still in wide use today. In the 1870s, deriving from uses in the California mining industry, Lester Allan Pelton developed the high efficiency Pelton wheel impulse turbine, which utilized hydropower from the high head streams characteristic of the mountainous California interior.

Hydraulic Power-pipe Networks

Hydraulic power networks used pipes to carrying pressurized water and transmit mechanical power from the source to end users. The power source was normally a head of water, which could also be assisted by a pump. These were extensive in Victorian cities in the United Kingdom. A hydraulic power network was also developed in Geneva, Switzerland. The world-famous Jet d'Eau was originally designed as the over-pressure relief valve for the network.

Compressed Air Hydro

Where there is a plentiful head of water it can be made to generate compressed air directly without moving parts. In these designs, a falling column of water is purposely mixed with air bubbles generated through turbulence or a venturi pressure reducer at the high level intake. This is allowed to fall down a shaft into a subterranean, high-roofed chamber where the now-compressed air separates from the water and becomes trapped. The height of the falling water column maintains compression of the air in the top of the chamber, while an outlet, submerged below the water level in the chamber allows water to flow back to the surface at a lower level than the intake. A separate outlet in the roof of the chamber supplies the compressed air. A facility on this principle was built on the Montreal River at Ragged Shutes near Cobalt, Ontario in 1910 and supplied 5,000 horsepower to nearby mines.

Hydropower Types

Hydropower is used primarily to generate electricity. Broad categories include:

- Conventional hydroelectric, referring to hydroelectric dams.

- Run-of-the-river hydroelectricity, which captures the kinetic energy in rivers or streams, without a large reservoir and sometimes without the use of dams.
- Small hydro projects are 10 megawatts or less and often have no artificial reservoirs.
- Micro hydro projects provide a few kilowatts to a few hundred kilowatts to isolated homes, villages, or small industries.
- Conduit hydroelectricity projects utilize water which has already been diverted for use elsewhere; in a municipal water system, for example.
- Pumped-storage hydroelectricity stores water pumped uphill into reservoirs during periods of low demand to be released for generation when demand is high or system generation is low.

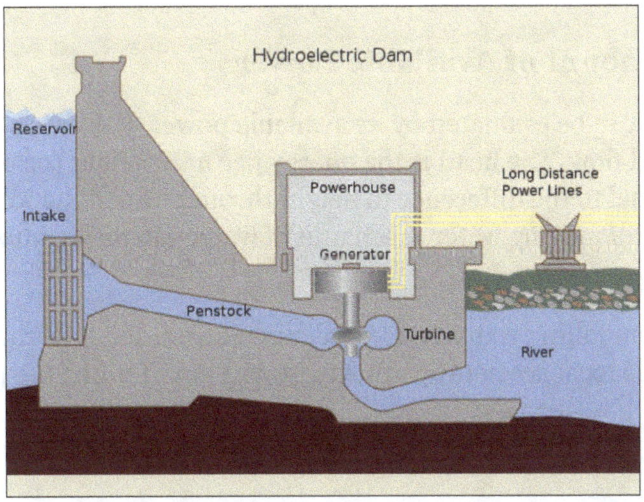

A conventional dammed-hydro facility (hydroelectric dam) is the most common type of hydroelectric power generation.

Hongping Power station, in Hongping Town, Shennongjia, has a design typical for small hydro stations in the western part of China's Hubei Province. Water comes from the mountain behind the station, through the black pipe seen in the photo

Chief Joseph Dam near Bridgeport, Washington, U.S., is a major run-of-the-river station without a sizeable reservoir.

Calculating The Amount of Available Power

A hydropower resource can be evaluated by its available power. Power is a function of the hydraulic head and rate of fluid flow. The head is the energy per unit weight (or unit mass) of water. The static head is proportional to the difference in height through which the water falls. Dynamic head is related to the velocity of moving water. Each unit of water can do an amount of work equal to its weight times the head.

The power available from falling water can be calculated from the flow rate and density of water, the height of fall, and the local acceleration due to gravity. In SI units, the power is:

$$P = \eta \rho Q g h$$

where,

- P is power in watts.
- η is the dimensionless efficiency of the turbine.
- ρ is the density of water in kilograms per cubic metre.
- Q is the flow in cubic metres per second.
- g is the acceleration due to gravity.
- h is the height difference between inlet and outlet in metres.

To illustrate, power is calculated for a turbine that is 85% efficient, with water at 1000 kg/cubic metre (62.5 pounds/cubic foot) and a flow rate of 80 cubic-meters/second (2800 cubic-feet/second), gravity of 9.81 metres per second squared and with a net head of 145 m (480 ft).

In SI units:

Power (W) $= 0.85 \times 1000 \times 80 \times 9.81 \times 145$ which gives 97 MW

In English units, the density is given in pounds per cubic foot so acceleration due to gravity is inherent in the unit of weight. A conversion factor is required to change from foot lbs/second to kilowatts:

$$\text{Power (W)} = 0.85 \times 62.5 \times 2800 \times 480 \times 1.356 \text{ which gives 97 MW (130,000 horsepower)}$$

Operators of hydroelectric stations will compare the total electrical energy produced with the theoretical potential energy of the water passing through the turbine to calculate efficiency. Procedures and definitions for calculation of efficiency are given in test codes such as ASME PTC 18 and IEC 60041. Field testing of turbines is used to validate the manufacturer's guaranteed efficiency. Detailed calculation of the efficiency of a hydropower turbine will account for the head lost due to flow friction in the power canal or penstock, rise in tail water level due to flow, the location of the station and effect of varying gravity, the temperature and barometric pressure of the air, the density of the water at ambient temperature, and the altitudes above sea level of the forebay and tailbay. For precise calculations, errors due to rounding and the number of significant digits of constants must be considered.

Some hydropower systems such as water wheels can draw power from the flow of a body of water without necessarily changing its height. In this case, the available power is the kinetic energy of the flowing water. Over-shot water wheels can efficiently capture both types of energy.

The water flow in a stream can vary widely from season to season. Development of a hydropower site requires analysis of flow records, sometimes spanning decades, to assess the reliable annual energy supply. Dams and reservoirs provide a more dependable source of power by smoothing seasonal changes in water flow. However reservoirs have significant environmental impact, as does alteration of naturally occurring stream flow. The design of dams must also account for the worst-case, "probable maximum flood" that can be expected at the site; a spillway is often included to bypass flood flows around the dam. A computer model of the hydraulic basin and rainfall and snowfall records are used to predict the maximum flood.

Hydropower Sustainability

As with other forms of economic activity, hydropower projects can have both a positive and a negative environmental and social impact, because the construction of a dam and power plant, along with the impounding of a reservoir, creates certain social and physical changes.

A number of tools have been developed to assist projects.

Most new hydropower project must undergo an Environmental and Social Impact Assessment. This provides a base line understand of the pre project conditions, estimates potential impacts and puts in place management plans to avoid, mitigate, or compensate for impacts.

The Hydropower Sustainability Assessment Protocol is another tool which can be used to promote and guide more sustainable hydropower projects. It is a methodology used to audit the performance of a hydropower project across more than twenty environmental, social, technical and economic topics. A Protocol assessment provides a rapid sustainability health check. It does not replace an environmental and social impact assessment (ESIA), which takes place over a much longer period of time, usually as a mandatory regulatory requirement.

The World Commission on Dams final report describes a framework for planning water and energy projects that is intended to protect dam-affected people and the environment, and ensure that the benefits from dams are more equitably distributed.

IFC's Environmental and Social Performance Standards define IFC clients' responsibilities for managing their environmental and social risks.

The World Bank's safeguard policies are used by the Bank to help identify, avoid, and minimize harms to people and the environment caused by investment projects.

The Equator Principles is a risk management framework, adopted by financial institutions, for determining, assessing and managing environmental and social risk in projects.

Hydroelectricity

Hydroelectricity is electricity produced from hydropower. In 2015 hydropower generated 16.6% of the world's total electricity and 70% of all renewable electricity, and was expected to increase about 3.1% each year for the next 25 years.

Hydropower is produced in 150 countries, with the Asia-Pacific region generating 33 percent of global hydropower in 2013. China is the largest hydroelectricity producer, with 920 TWh of production in 2013, representing 16.9 percent of domestic electricity use.

The cost of hydroelectricity is relatively low, making it a competitive source of renewable electricity. The hydro station consumes no water, unlike coal or gas plants. The average cost of electricity from a hydro station larger than 10 megawatts is 3 to 5 U.S. cents per kilowatt-hour. With a dam and reservoir it is also a flexible source of electricity since the amount produced by the station can be changed up or down very quickly to adapt to changing energy demands. Once a hydroelectric complex is constructed, the project produces no direct waste, and has a considerably lower output level of greenhouse gases than fossil fuel powered energy plants.

History

Museum Hydroelectric power plant "Under the Town" in Serbia, built in 1900.

Hydropower has been used since ancient times to grind flour and perform other tasks. In the mid-1770s, French engineer Bernard Forest de Bélidor published *Architecture Hydraulique* which described vertical- and horizontal-axis hydraulic machines. By the late 19th century, the electrical generator was developed and could now be coupled with hydraulics. The growing demand for the Industrial Revolution would drive development as well. In 1878 the world's first hydroelectric power scheme was developed at Cragside in Northumberland, England by William George Armstrong. It was used to power a single arc lamp in his art gallery. The old Schoelkopf Power Station No. 1 near Niagara Falls in the U.S. side began to produce electricity in 1881. The first Edison hydroelectric power station, the Vulcan Street Plant, began operating September 30, 1882, in Appleton, Wisconsin, with an output of about 12.5 kilowatts. By 1886 there were 45 hydroelectric power stations in the U.S. and Canada. By 1889 there were 200 in the U.S. alone.

At the beginning of the 20th century, many small hydroelectric power stations were being constructed by commercial companies in mountains near metropolitan areas. Grenoble, France held the International Exhibition of Hydropower and Tourism with over one million visitors. By 1920 as 40% of the power produced in the United States was hydroelectric, the Federal Power Act was enacted into law. The Act created the Federal Power Commission to regulate hydroelectric power stations on federal land and water. As the power stations became larger, their associated dams developed additional purposes to include flood control, irrigation and navigation. Federal funding became necessary for large-scale development and federally owned corporations, such as the Tennessee Valley Authority (1933) and the Bonneville Power Administration (1937) were created. Additionally, the Bureau of Reclamation which had begun a series of western U.S. irrigation projects in the early 20th century was now constructing large hydroelectric projects such as the 1928 Hoover Dam. The U.S. Army Corps of Engineers was also involved in hydroelectric development, completing the Bonneville Dam in 1937 and being recognized by the Flood Control Act of 1936 as the premier federal flood control agency.

Hydroelectric power stations continued to become larger throughout the 20th century. Hydropower was referred to as *white coal* for its power and plenty. Hoover Dam's initial 1,345 MW power station was the world's largest hydroelectric power station in 1936; it was eclipsed by the 6809 MW Grand Coulee Dam in 1942. The Itaipu Dam opened in 1984 in South America as the largest, producing 14,000 MW but was surpassed in 2008 by the Three Gorges Dam in China at 22,500 MW. Hydroelectricity would eventually supply some countries, including Norway, Democratic Republic of the Congo, Paraguay and Brazil, with over 85% of their electricity. The United States currently has over 2,000 hydroelectric power stations that supply 6.4% of its total electrical production output, which is 49% of its renewable electricity.

Generating Methods

Turbine row at El Nihuil II Power Station in Mendoza, Argentina.

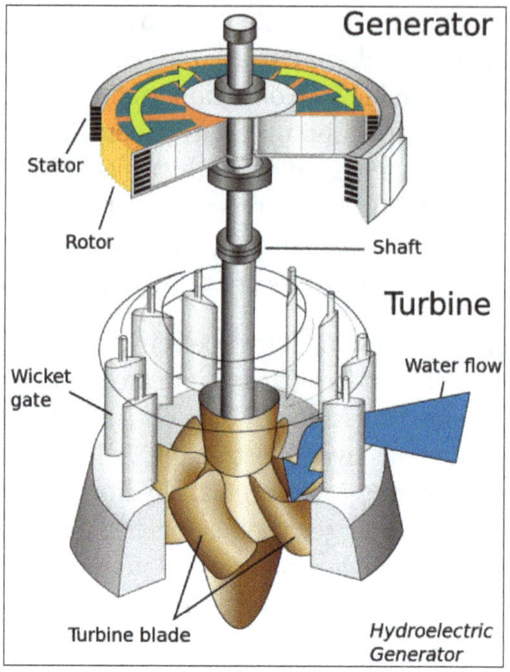

A typical turbine and generator.

Conventional (Dams)

Most hydroelectric power comes from the potential energy of dammed water driving a water turbine and generator. The power extracted from the water depends on the volume and on the difference in height between the source and the water's outflow. This height difference is called the head. A large pipe (the "penstock") delivers water from the reservoir to the turbine.

Pumped-Storage

This method produces electricity to supply high peak demands by moving water between reservoirs at different elevations. At times of low electrical demand, the excess generation capacity is used to pump water into the higher reservoir. When the demand becomes greater, water is released back into the lower reservoir through a turbine. Pumped-storage schemes currently provide the most commercially important means of large-scale grid energy storage and improve the daily capacity factor of the generation system. Pumped storage is not an energy source, and appears as a negative number in listings.

Run-of-the-River

Run-of-the-river hydroelectric stations are those with small or no reservoir capacity, so that only the water coming from upstream is available for generation at that moment, and any oversupply must pass unused. A constant supply of water from a lake or existing reservoir upstream is a significant advantage in choosing sites for run-of-the-river. In the United States, run of the river hydropower could potentially provide 60,000 megawatts (80,000,000 hp) (about 13.7% of total use in 2011 if continuously available).

Tide

A tidal power station makes use of the daily rise and fall of ocean water due to tides; such sources are highly predictable, and if conditions permit construction of reservoirs, can also be dispatchable to generate power during high demand periods. Less common types of hydro schemes use water's kinetic energy or undammed sources such as undershot water wheels. Tidal power is viable in a relatively small number of locations around the world. In Great Britain, there are eight sites that could be developed, which have the potential to generate 20% of the electricity used in 2012.

Sizes, Types and Capacities of Hydroelectric Facilities

Large Facilities

Large-scale hydroelectric power stations are more commonly seen as the largest power producing facilities in the world, with some hydroelectric facilities capable of generating more than double the installed capacities of the current largest nuclear power stations.

Although no official definition exists for the capacity range of large hydroelectric power stations, facilities from over a few hundred megawatts are generally considered large hydroelectric facilities.

Currently, only four facilities over 10 GW (10,000 MW) are in operation worldwide, see table below.

Rank	Station	Country	Location	Capacity (MW)
1.	Three Gorges Dam	China	30°49′15″N 111°00′08″E30.82083°N 111.00222°E	22,500
2.	Itaipu Dam	Brazil Paraguay	25°24′31″S 54°35′21″W25.40861°S 54.58917°W	14,000
3.	Xiluodu Dam	China	28°15′35″N 103°38′58″E28.25972°N 103.64944°E	13,860
4.	Guri Dam	Venezuela	07°45′59″N 62°59′57″W7.76639°N 62.99917°W	10,200

Panoramic view of the Itaipu Dam, with the spillways (closed at the time of the photo) on the left. In 1994, the American Society of Civil Engineers elected the Itaipu Dam as one of the seven modern Wonders of the World.

Small

Small hydro is the development of hydroelectric power on a scale serving a small community or industrial plant. The definition of a small hydro project varies but a generating capacity of up to 10 megawatts (MW) is generally accepted as the upper limit of what can be termed small hydro. This may be stretched to 25 MW and 30 MW in Canada and the United States. Small-scale hydro-electricity production grew by 28% during 2008 from 2005, raising the total world small-hydro capacity to 85 GW. Over 70% of this was in China (65 GW), followed by Japan (3.5 GW), the United States (3 GW), and India (2 GW).

A micro-hydro facility in Vietnam.

Small hydro stations may be connected to conventional electrical distribution networks as a source of low-cost renewable energy. Alternatively, small hydro projects may be built in isolated areas that would be uneconomic to serve from a network, or in areas where there is no national electrical distribution network. Since small hydro projects usually have minimal reservoirs and civil construction work, they are seen as having a relatively low environmental impact compared to large hydro. This decreased environmental impact depends strongly on the balance between stream flow and power production.

Pico hydroelectricity in Mondulkiri, Cambodia.

Micro

Micro hydro is a term used for hydroelectric power installations that typically produce up to 100 kW of power. These installations can provide power to an isolated home or small community, or are sometimes connected to electric power networks. There are many of these installations around the world, particularly in developing nations as they can provide an economical source of energy without purchase of fuel. Micro hydro systems complement photovoltaic solar energy systems because in many areas, water flow, and thus available hydro power, is highest in the winter when solar energy is at a minimum.

Pico

Pico hydro is a term used for hydroelectric power generation of under 5 kW. It is useful in small, remote communities that require only a small amount of electricity. For example, to power one or two fluorescent light bulbs and a TV or radio for a few homes. Even smaller turbines of 200-300W may power a single home in a developing country with a drop of only 1 m (3 ft). A Pico-hydro setup is typically run-of-the-river, meaning that dams are not used, but rather pipes divert some of the flow, drop this down a gradient, and through the turbine before returning it to the stream.

Underground

An underground power station is generally used at large facilities and makes use of a large natural height difference between two waterways, such as a waterfall or mountain lake. An underground tunnel is constructed to take water from the high reservoir to the generating hall built in an underground cavern near the lowest point of the water tunnel and a horizontal tailrace taking water away to the lower outlet waterway.

Measurement of the tailrace and forebay rates at the Limestone Generating Station in Manitoba, Canada.

Calculating Available Power

A simple formula for approximating electric power production at a hydroelectric station is: , where

- Is Power in watts,
- Is the density of water (~1000 kg/m^3),
- Is height in meters,
- Is flow rate in cubic meters per second,
- Is acceleration due to gravity of 9.8 m/s^2,
- Is a coefficient of efficiency ranging from 0 to 1. Efficiency is often higher (that is, closer to 1) with larger and more modern turbines.

Annual electric energy production depends on the available water supply. In some installations, the water flow rate can vary by a factor of 10:1 over the course of a year.

Properties

Advantages

The Ffestiniog Power Station can generate 360 MW of electricity within 60 seconds of the demand arising.

Flexibility

Hydropower is a flexible source of electricity since stations can be ramped up and down very quickly to adapt to changing energy demands. Hydro turbines have a start-up time of the order of a few minutes. It takes around 60 to 90 seconds to bring a unit from cold start-up to full load; this is much shorter than for gas turbines or steam plants. Power generation can also be decreased quickly when there is a surplus power generation. Hence the limited capacity of hydropower units is not generally used to produce base power except for vacating the flood pool or meeting downstream needs. Instead, it serves as backup for non-hydro generators.

Low Power Costs

The major advantage of hydroelectricity is elimination of the cost of fuel. The cost of operating a hydroelectric station is nearly immune to increases in the cost of fossil fuels such as oil, natural gas or coal, and no imports are needed. The average cost of electricity from a hydro station larger than 10 megawatts is 3 to 5 U.S. cents per kilowatt-hour.

Hydroelectric stations have long economic lives, with some plants still in service after 50–100 years. Operating labor cost is also usually low, as plants are automated and have few personnel on site during normal operation.

Where a dam serves multiple purposes, a hydroelectric station may be added with relatively low construction cost, providing a useful revenue stream to offset the costs of dam operation. It has been calculated that the sale of electricity from the Three Gorges Dam will cover the construction costs after 5 to 8 years of full generation. Additionally, some data shows that in most countries large hydropower dams will be too costly and take too long to build to deliver a positive risk adjusted return, unless appropriate risk management measures are put in place.

Suitability for Industrial Applications

While many hydroelectric projects supply public electricity networks, some are created to serve specific industrial enterprises. Dedicated hydroelectric projects are often built to provide the substantial amounts of electricity needed for aluminium electrolytic plants, for example. The Grand Coulee Dam switched to support Alcoa aluminium in Bellingham, Washington, United States for American World War II airplanes before it was allowed to provide irrigation and power to citizens (in addition to aluminium power) after the war. In Suriname, the Brokopondo Reservoir was constructed to provide electricity for the Alcoa aluminium industry. New Zealand's Manapouri Power Station was constructed to supply electricity to the aluminium smelter at Tiwai Point.

Reduced CO_2 emissions

Since hydroelectric dams do not burn fossil fuels, they do not directly produce carbon dioxide. While some carbon dioxide is produced during manufacture and construction of the project, this is a tiny fraction of the operating emissions of equivalent fossil-fuel electricity generation. One measurement of greenhouse gas related and other externality comparison between energy sources can be found in the ExternE project by the Paul Scherrer Institute and the University of Stuttgart which was funded by the European Commission. According to that study, hydroelectricity produces the least amount of greenhouse gases and externality of any energy source. Coming in second place was wind, third was nuclear energy, and fourth was solar photovoltaic. The low greenhouse gas impact of hydroelectricity is found especially in temperate climates. The above study was for local energy in Europe; presumably similar conditions prevail in North America and Northern Asia, which all see a regular, natural freeze/thaw cycle (with associated seasonal plant decay and regrowth). Greater greenhouse gas emission impacts are found in the tropical regions because the reservoirs of power stations in tropical regions produce a larger amount of methane than those in temperate areas.

Other Uses of The Reservoir

Reservoirs created by hydroelectric schemes often provide facilities for water sports, and become tourist attractions themselves. In some countries, aquaculture in reservoirs is common. Multi-use dams installed for irrigation support agriculture with a relatively constant water supply. Large hydro dams can control floods, which would otherwise affect people living downstream of the project.

Disadvantages

Large reservoirs associated with traditional hydroelectric power stations result in submersion of extensive areas upstream of the dams, sometimes destroying biologically rich and productive lowland and riverine valley forests, marshland and grasslands. Damming interrupts the flow of rivers and can harm local ecosystems, and building large dams and reservoirs often involves displacing people and wildlife. The loss of land is often exacerbated by habitat fragmentation of surrounding areas caused by the reservoir.

Hydroelectric projects can be disruptive to surrounding aquatic ecosystems both upstream and downstream of the plant site. Generation of hydroelectric power changes the downstream river environment. Water exiting a turbine usually contains very little suspended sediment, which can lead to scouring of river beds and loss of riverbanks. Since turbine gates are often opened intermittently, rapid or even daily fluctuations in river flow are observed.

Hydroelectric power stations that use dams would submerge large areas of land due to the requirement of a reservoir. Merowe Dam in Sudan.

Siltation and Flow Shortage

When water flows it has the ability to transport particles heavier than itself downstream. This has a negative effect on dams and subsequently their power stations, particularly those on rivers or within catchment areas with high siltation. Siltation can fill a reservoir and reduce its capacity to control floods along with causing additional horizontal pressure on the upstream portion of the dam. Eventually, some reservoirs can become full of sediment and useless or over-top during a flood and fail.

Changes in the amount of river flow will correlate with the amount of energy produced by a dam. Lower river flows will reduce the amount of live storage in a reservoir therefore reducing the amount of water that can be used for hydroelectricity. The result of diminished river flow can be power shortages in areas that depend heavily on hydroelectric power. The risk of flow shortage may increase as a result of climate change. One study from the Colorado River in the United States suggest that modest climate changes, such as an increase in temperature in 2 degree Celsius resulting in a 10% decline in precipitation, might reduce river run-off by up to 40%. Brazil in particular is vulnerable due to its heavy reliance on hydroelectricity, as increasing temperatures, lower water flow and alterations in the rainfall regime, could reduce total energy production by 7% annually by the end of the century.

Methane Emissions (From Reservoirs)

Lower positive impacts are found in the tropical regions, as it has been noted that the reservoirs of power plants in tropical regions produce substantial amounts of methane. This is due to plant material in flooded areas decaying in an anaerobic environment, and forming methane, a greenhouse gas. According to the World Commission on Dams report, where the reservoir is large compared to the generating capacity (less than 100 watts per square metre of surface area) and no clearing of the forests in the area was undertaken prior to impoundment of the reservoir, greenhouse gas emissions from the reservoir may be higher than those of a conventional oil-fired thermal generation plant.

In boreal reservoirs of Canada and Northern Europe, however, greenhouse gas emissions are typically only 2% to 8% of any kind of conventional fossil-fuel thermal generation. A new class of underwater logging operation that targets drowned forests can mitigate the effect of forest decay.

The Hoover Dam in the United States is a large conventional dammed-hydro facility, with an installed capacity of 2,080 MW.

Relocation

Another disadvantage of hydroelectric dams is the need to relocate the people living where the reservoirs are planned. In 2000, the World Commission on Dams estimated that dams had physically displaced 40-80 million people worldwide.

Failure Risks

Because large conventional dammed-hydro facilities hold back large volumes of water, a failure due to poor construction, natural disasters or sabotage can be catastrophic to downriver settlements and infrastructure. Dam failures have been some of the largest man-made disasters in history.

During Typhoon Nina in 1975 Banqiao Dam failed in Southern China when more than a year's worth of rain fell within 24 hours. The resulting flood resulted in the deaths of 26,000 people, and another 145,000 from epidemics. Millions were left homeless. Also, the creation of a dam in a geologically inappropriate location may cause disasters such as 1963 disaster at Vajont Dam in Italy, where almost 2,000 people died.

The Malpasset Dam failure in Fréjus on the French Riviera (Côte d'Azur), southern France, collapsed on December 2, 1959, killing 423 people in the resulting flood.

Smaller dams and micro hydro facilities create less risk, but can form continuing hazards even after being decommissioned. For example, the small Kelly Barnes Dam failed in 1967, causing 39 deaths with the Toccoa Flood, ten years after its power station was decommissioned the earthen embankment dam failed.

Comparison with Other Methods of Power Generation

Hydroelectricity eliminates the flue gas emissions from fossil fuel combustion, including pollutants such as sulfur dioxide, nitric oxide, carbon monoxide, dust, and mercury in the coal. Hydroelectricity also avoids the hazards of coal mining and the indirect health effects of coal emissions. Compared to nuclear power, hydroelectricity construction requires altering large areas of the environment while a nuclear power station has a small footprint, and hydro-powerstation failures have caused tens of thousands of more deaths than any nuclear station failure. The creation of Garrison Dam, for example, required Native American land to create Lake Sakakawea, which has a shoreline of 1,320 miles, and caused the inhabitants to sell 94% of their arable land for $7.5 million in 1949.

Compared to wind farms, hydroelectricity power stations have a more predictable load factor. If the project has a storage reservoir, it can generate power when needed. Hydroelectric stations can be easily regulated to follow variations in power demand.

World Hydroelectric Capacity

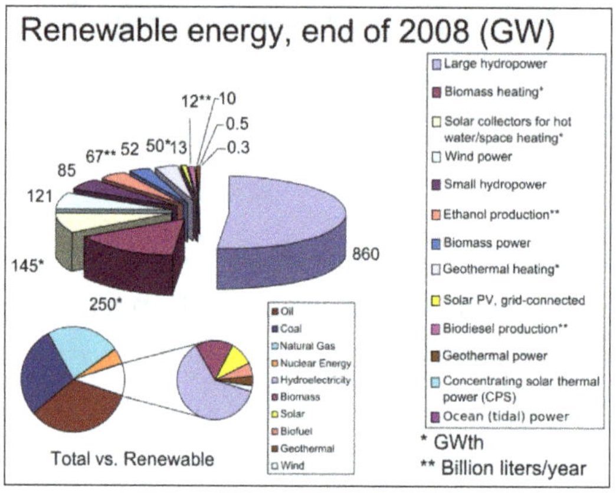

World renewable energy share (2008).

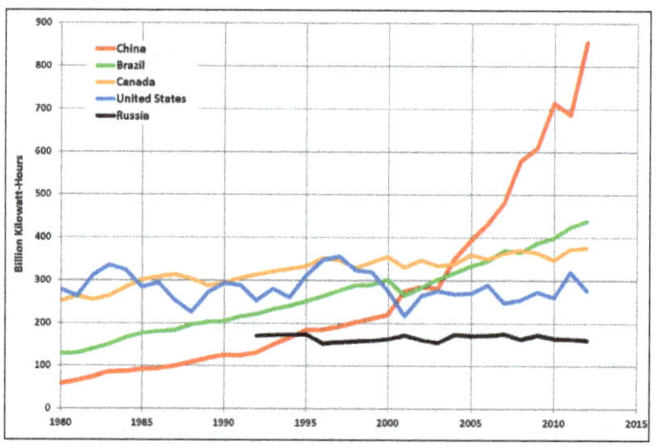

Trends in the top five hydroelectricity-producing countries.

The ranking of hydro-electric capacity is either by actual annual energy production or by installed capacity power rating. In 2015 hydropower generated 16.6% of the worlds total electricity and 70%

of all renewable electricity. Hydropower is produced in 150 countries, with the Asia-Pacific region generated 32 percent of global hydropower in 2010. China is the largest hydroelectricity producer, with 721 terawatt-hours of production in 2010, representing around 17 percent of domestic electricity use. Brazil, Canada, New Zealand, Norway, Paraguay, Austria, Switzerland, and Venezuela have a majority of the internal electric energy production from hydroelectric power. Paraguay produces 100% of its electricity from hydroelectric dams, and exports 90% of its production to Brazil and to Argentina. Norway produces 98–99% of its electricity from hydroelectric sources.

A hydro-electric station rarely operates at its full power rating over a full year; the ratio between annual average power and installed capacity rating is the capacity factor. The installed capacity is the sum of all generator nameplate power ratings.

Ten of the largest hydroelectric producers as at 2013.

Country	Annual hydroelectric production (TWh)	Installed capacity (GW)	Capacity factor	% of total production
China	920	194	0.37	16.9%
Canada	392	76	0.59	60.1%
Brazil	391	86	0.56	68.6%
United States	290	102	0.42	6.7
Russia	183	50	0.42	17.3%
India	142	40	0.43	11.9%
Norway	129	31	0.49	96.1%
Japan	85	49	0.37	8.1%
Venezuela	84	15	0.67	67.8%
France	76	25	0.46	13.2%

Major Projects Under Construction

Name	Maximum Capacity	Country	Construction started	Scheduled completion	Comments
Belo Monte Dam	11,181 MW	Brazil	March, 2011	2015	Preliminary construction underway. Construction suspended 14 days by court order Aug 2012
Siang Upper HE Project	11,000 MW	India	April, 2009	2024	Multi-phase construction over a period of 15 years. Construction was delayed due to dispute with China.
Tasang Dam	7,110 MW	Burma	March, 2007	2022	Controversial 228 meter tall dam with capacity to produce 35,446 GWh annually.

Xiangjiaba Dam	6,400 MW	China	November 26, 2006	2015	The last generator was commissioned on July 9, 2014
Grand Ethiopian Renaissance Dam	6,000 MW	Ethiopia	2011	2017	Located in the upper Nile Basin, drawing complaint from Egypt
Nuozhadu Dam	5,850 MW	China	2006	2017	
Jinping 2 Hydropower Station	4,800 MW	China	January 30, 2007	2014	To build this dam, 23 families and 129 local residents need to be moved. It works with Jinping 1 Hydropower Station as a group.
Diamer-Bhasha Dam	4,500 MW	Pakistan	October 18, 2011	2023	
Jinping 1 Hydropower Station	3,600 MW	China	November 11, 2005	2014	The sixth and final generator was commissioned on 15 July 2014
Jirau Power Station	3,300 MW	Brazil	2008	2013	Construction halted in March 2011 due to worker riots.
Guanyinyan Dam	3,000 MW	China	2008	2015	Construction of the roads and spillway started.
Lianghekou Dam	3,000 MW	China	2014	2023	
Dagangshan Dam	2,600 MW	China	August 15, 2008	2016	
Liyuan Dam	2,400 MW	China	2008	2013	
Tocoma Dam Bolívar State	2,160 MW	Venezuela	2004	2014	This power station would be the last development in the Low Caroni Basin, bringing the total to six power stations on the same river, including the 10,000MW Guri Dam.
Ludila Dam	2,100 MW	China	2007	2015	Brief construction halt in 2009 for environmental assessment.
Shuangjiangkou Dam	2,000 MW	China	December, 2007	2018	The dam will be 312 m high.
Ahai Dam	2,000 MW	China	July 27, 2006	2015	
Teles Pires Dam	1,820 MW	Brazil	2011	2015	
Site C Dam	1,100 MW	Canada	2015	2024	First large dam in western Canada since 1984
Lower Subansiri Dam	2,000 MW	India	2007	2016	

Water Turbine

A water turbine is a rotary machine that converts kinetic energy and potential energy of water into mechanical work.

The runner of the small water turbine.

Water turbines were developed in the 19th century and were widely used for industrial power prior to electrical grids. Now they are mostly used for electric power generation. Water turbines are mostly found in dams to generate electric power from water kinetic energy.

History

Water wheels have been used for hundreds of years for industrial power. Their main shortcoming is size, which limits the flow rate and head that can be harnessed. The migration from water wheels to modern turbines took about one hundred years. Development occurred during the Industrial revolution, using scientific principles and methods. They also made extensive use of new materials and manufacturing methods developed at the time.

The construction of a Ganz water Turbo Generator in Budapest in 1886.

Swirl

The word turbine was introduced by the French engineer Claude Burdin in the early 19th century and is derived from the Latin word for "whirling" or a "vortex". The main difference between early water turbines and water wheels is a swirl component of the water which passes energy to a spinning rotor. This additional component of motion allowed the turbine to be smaller than a water wheel of the same power. They could process more water by spinning faster and could harness much greater heads. (Later, impulse turbines were developed which didn't use swirl).

Timeline

The earliest known water turbines date to the Roman Empire. Two helix-turbine mill sites of almost identical design were found at Chemtou and Testour, modern-day Tunisia, dating to the late 3rd or early 4th century AD. The horizontal water wheel with angled blades was installed at the bottom of a water-filled, circular shaft. The water from the mill-race entered the pit tangentially, creating a swirling water column which made the fully submerged wheel act like a true turbine.

Roman turbine mill at Chemtou, Tunisia. The tangential water inflow of the millrace made the submerged horizontal wheel in the shaft turn like a true turbine.

Fausto Veranzio in his book Machinae Novae (1595) described a vertical axis mill with a rotor similar to that of a Francis turbine.

A Francis turbine runner, rated at nearly one million hp (750 MW), being installed at the Grand Coulee Dam, United States.

Johann Segner developed a reactive water turbine (Segner wheel) in the mid-18th century in Kingdom of Hungary. It had a horizontal axis and was a precursor to modern water turbines. It is a very simple machine that is still produced today for use in small hydro sites. Segner worked with Euler on some of the early mathematical theories of turbine design. In the 18th century, a Dr. Barker invented a similar reaction hydraulic turbine that became popular as a lecture-hall demonstration. The only known surviving example of this type of engine used in power production, dating from 1851, is found at Hacienda Buena Vista in Ponce, Puerto Rico.

A propeller-type runner rated 28,000 hp (21 MW).

In 1820, Jean-Victor Poncelet developed an inward-flow turbine.

In 1826, Benoît Fourneyron developed an outward-flow turbine. This was an efficient machine (~80%) that sent water through a runner with blades curved in one dimension. The stationary outlet also had curved guides.

In 1844, Uriah A. Boyden developed an outward flow turbine that improved on the performance of the Fourneyron turbine. Its runner shape was similar to that of a Francis turbine.

In 1849, James B. Francis improved the inward flow reaction turbine to over 90% efficiency. He also conducted sophisticated tests and developed engineering methods for water turbine design. The Francis turbine, named for him, is the first modern water turbine. It is still the most widely used water turbine in the world today. The Francis turbine is also called a radial flow turbine, since water flows from the outer circumference towards the centre of runner.

Inward flow water turbines have a better mechanical arrangement and all modern reaction water turbines are of this design. As the water swirls inward, it accelerates, and transfers energy to the runner. Water pressure decreases to atmospheric, or in some cases subatmospheric, as the water passes through the turbine blades and loses energy.

Around 1890, the modern fluid bearing was invented, now universally used to support heavy water turbine spindles. As of 2002, fluid bearings appear to have a mean time between failures of more than 1300 years.

Around 1913, Viktor Kaplan created the Kaplan turbine, a propeller-type machine. It was an evolution of the Francis turbine but revolutionized the ability to develop low-head hydro sites.

New Concept

All common water machines until the late 19th century (including water wheels) were basically reaction machines; water *pressure* head acted on the machine and produced work. A reaction turbine needs to fully contain the water during energy transfer.

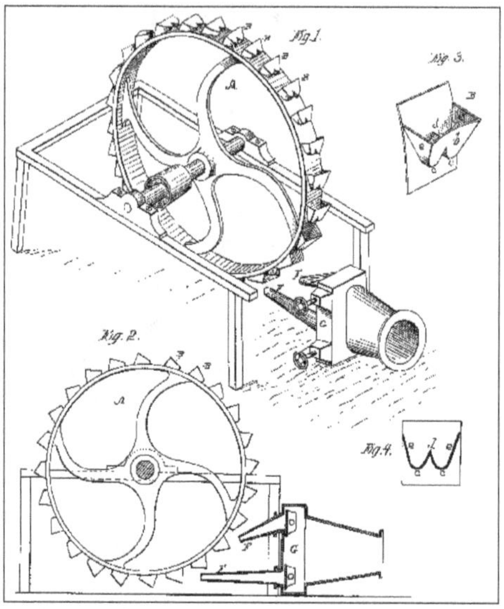

Figure from Pelton's original patent (October 1880).

In 1866, California millwright Samuel Knight invented a machine that took the impulse system to a new level. Inspired by the high pressure jet systems used in hydraulic mining in the gold fields, Knight developed a bucketed wheel which captured the energy of a free jet, which had converted a high head (hundreds of vertical feet in a pipe or penstock) of water to kinetic energy. This is called an impulse or tangential turbine. The water's velocity, roughly twice the velocity of the bucket periphery, does a u-turn in the bucket and drops out of the runner at low velocity.

In 1879, Lester Pelton, experimenting with a Knight Wheel, developed a Pelton wheel (double bucket design), which exhausted the water to the side, eliminating some energy loss of the Knight wheel which exhausted some water back against the center of the wheel. In about 1895, William Doble improved on Pelton's half-cylindrical bucket form with an elliptical bucket that included a cut in it to allow the jet a cleaner bucket entry. This is the modern form of the Pelton turbine which today achieves up to 92% efficiency. Pelton had been quite an effective promoter of his design and although Doble took over the Pelton company he did not change the name to Doble because it had brand name recognition.

Turgo and cross-flow turbines were later impulse designs.

Theory of Operation

Flowing water is directed on to the blades of a turbine runner, creating a force on the blades. Since the runner is spinning, the force acts through a distance (force acting through a distance is the definition of work). In this way, energy is transferred from the water flow to the turbine

Water turbines are divided into two groups; reaction turbines and impulse turbines.

The precise shape of water turbine blades is a function of the supply pressure of water, and the type of impeller selected.

Reaction Turbines

Reaction turbines are acted on by water, which changes pressure as it moves through the turbine and gives up its energy. They must be encased to contain the water pressure (or suction), or they must be fully submerged in the water flow.

Newton's third law describes the transfer of energy for reaction turbines.

Most water turbines in use are reaction turbines and are used in low (<30 m or 100 ft) and medium (30–300 m or 100–1,000 ft) head applications. In reaction turbine pressure drop occurs in both fixed and moving blades. It is largely used in dam and large power plants.

Impulse Turbines

Impulse turbines change the velocity of a water jet. The jet pushes on the turbine's curved blades which changes the direction of the flow. The resulting change in momentum (impulse) causes a force on the turbine blades. Since the turbine is spinning, the force acts through a distance (work) and the diverted water flow is left with diminished energy. An impulse turbine is one which the pressure of the fluid flowing over the rotor blades is constant and all the work output is due to the change in kinetic energy of the fluid.

Prior to hitting the turbine blades, the water's pressure (potential energy) is converted to kinetic energy by a nozzle and focused on the turbine. No pressure change occurs at the turbine blades, and the turbine doesn't require a housing for operation.

Newton's second law describes the transfer of energy for impulse turbines.

Impulse turbines are often used in very high (>300m/1000 ft) head applications.

Power

The power available in a stream of water is; $P = \eta \cdot \rho \cdot g \cdot h \cdot \dot{q}$,

- P = power (J/s or watts),
- η = turbine efficiency,
- ρ = density of water (kg/m^3),
- g = acceleration of gravity (9.81 m/s^2),
- h = head (m). For still water, this is the difference in height between the inlet and outlet surfaces. Moving water has an additional component added to account for the kinetic energy of the flow. The total head equals the *pressure head* plus *velocity head.*
- $\dot{q}$ = flow rate (m^3/s).

Pumped-storage Hydroelectricity

Some water turbines are designed for pumped-storage hydroelectricity. They can reverse flow and operate as a pump to fill a high reservoir during off-peak electrical hours, and then revert to a water turbine for power generation during peak electrical demand. This type of turbine is usually a Deriaz or Francis turbine in design.

Efficiency

Large modern water turbines operate at mechanical efficiencies greater than 90%.

Types of Water Turbines

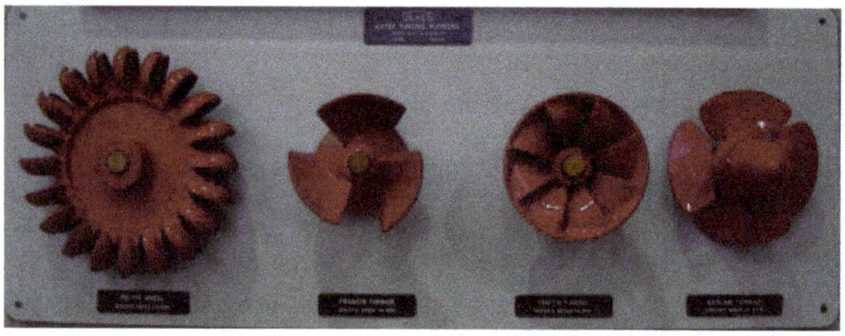

Various types of water turbine runners. From left to right: Pelton wheel, two types of Francis turbine and Kaplan turbine.

Reaction turbines:

- VLH turbine,
- Francis turbine,
- Kaplan turbine,
- Tyson turbine,
- Gorlov helical turbine.

Impulse turbine:

- Water wheel,
- Pelton wheel,
- Turgo turbine,
- Cross-flow turbine (also known as the Bánki-Michell turbine, or Ossberger turbine).
- Jonval turbine,
- Reverse overshot water-wheel,
- Screw turbine,
- Barkh Turbine,

Design and Application

Small turbines (mostly under 10 MW) may have horizontal shafts, and even fairly large bulb-type turbines up to 100 MW or so may be horizontal. Very large Francis and Kaplan machines usually have vertical shafts because this makes best use of the available head, and makes installation of a generator more economical. Pelton wheels may be either vertical or horizontal shaft machines because the size of the machine is so much less than the available head. Some impulse turbines use multiple jets per runner to balance shaft thrust. This also allows for the use of a smaller turbine runner, which can decrease costs and mechanical losses.

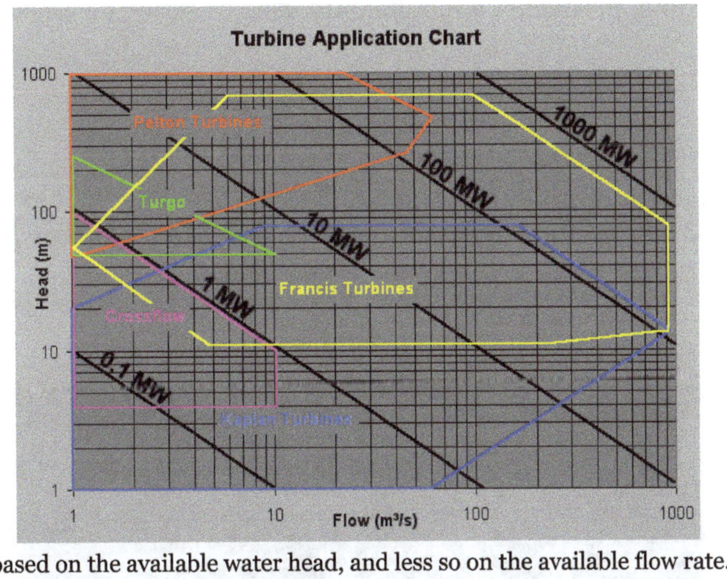

Turbine selection is based on the available water head, and less so on the available flow rate. In general, impulse turbines are used for high head sites, and reaction turbines are used for low head sites. Kaplan turbines with adjustable blade pitch are well-adapted to wide ranges of flow or head conditions, since their peak efficiency can be achieved over a wide range of flow conditions.

Typical Range of Heads

• Water wheel	$0.2 < H < 4$ (H = head in m)
• Screw turbine	$1 < H < 10$
• VLH turbine	$1.5 < H < 4.5$
• Kaplan turbine	$20 < H < 40$
• Francis turbine	$40 < H < 600$
• Pelton wheel	$50 < H < 1300$
• Turgo turbine	$50 < H < 250$

Specific Speed

The specific speed of a turbine characterizes the turbine's shape in a way that is not related to its size. This allows a new turbine design to be scaled from an existing design of known performance. The specific speed is also the main criteria for matching a specific hydro site with the correct turbine type. The specific speed is the speed with which the turbine turns for a particular discharge Q, with unit head and thereby is able to produce unit power.

Affinity Laws

Affinity laws allow the output of a turbine to be predicted based on model tests. A miniature replica of a proposed design, about one foot (0.3 m) in diameter, can be tested and the laboratory measurements applied to the final application with high confidence. Affinity laws are derived by requiring similitude between the test model and the application.

Flow through the turbine is controlled either by a large valve or by wicket gates arranged around the outside of the turbine runner. Differential head and flow can be plotted for a number of different values of gate opening, producing a hill diagram used to show the efficiency of the turbine at varying conditions.

Runaway Speed

The runaway speed of a water turbine is its speed at full flow, and no shaft load. The turbine will be designed to survive the mechanical forces of this speed. The manufacturer will supply the runaway speed rating.

Control Systems

Operation of a flyball governor to control speeds of a water turbine.

Different designs of governors have been used since the mid-19th century to control the speeds of the water turbines. A variety of flyball systems, or first-generation governors, were used during the first 100 years of water turbine speed controls. In early flyball systems, the flyball component countered by a spring acted directly to the valve of the turbine or the wicket gate to control the amount of water that enters the turbines. Newer systems with mechanical governors started around 1880. An early mechanical governors is a servomechanism that comprises a series of gears that use the turbine's speed to drive the flyball and turbine's power to drive the control mechanism. The mechanical governors were continued to be enhanced in power amplification through the use of gears and the dynamic behavior. By 1930, the mechanical governors had many parameters that could be set on the feedback system for precise controls. In the later part of the twentieth century, electronic governors and digital systems started to replace the mechanical governors. In the electronic governors, also known as second-generation governors, the flyball was replaced by rotational speed sensor but the controls were still done through analog systems. In the modern systems, also known as third-generation governors, the controls are performed digitally by algorithms that are programmed to the computer of the governor.

Turbine Blade Materials

Given that the turbine blades in a water turbine are constantly exposed to water and dynamic forces, they need to have high corrosion resistance and strength. The most common material used in overlays on carbon steel runners in water turbines are austenitic steel alloys that have 17% to 20% chromium to increase stability of the film which improves aqueous corrosion resistance. The chromium content in these steel alloys exceed the minimum of 12% chromium required to exhibit some atmospheric corrosion resistance. Having a higher chromium concentration in the steel alloys allows for a much longer lifespan of the turbine blades. Currently, the blades are made of martensitic stainless steels which have high strength compared to austenitic stainless steels by a factor of 2. Besides corrosion resistance and strength as the criteria for material selection, weld-ability and density of the turbine blade. Greater weld-ability allows for easier repair of the turbine blades. This also allows for higher weld quality which results in a better repair. Selecting a material with low density is important to achieve higher efficiency because the lighter blades rotate more easily. The most common material used in Kaplan Turbine blades are stainless steel alloys (SS). The different alloys used are SS(16Cr-5Ni), SS(13Cr-4Ni), SS(13Cr-1Ni). The martensitic stainless steel alloys have high strength, thinner sections than standard carbon steel, and reduced mass that enhances the hydrodynamic flow conditions and efficiency of the water turbine. The SS(13Cr-4Ni) has been shown to have improved erosion resistance at all angles of attack through the process of laser hardening. It is important to minimize erosion in order to maintain high efficiencies because erosion negatively impacts the hydraulic profile of the blades which reduces the relative ease to rotate.

Maintenance

Turbines are designed to run for decades with very little maintenance of the main elements; overhaul intervals are on the order of several years. Maintenance of the runners and parts exposed to water include removal, inspection, and repair of worn parts.

A Francis turbine at the end of its life showing pitting corrosion, fatigue cracking and a catastrophic failure. Earlier repair jobs that used stainless steel weld rods are visible.

Normal wear and tear includes pitting corrosion from cavitation, fatigue cracking, and abrasion from suspended solids in the water. Steel elements are repaired by welding, usually with stainless steel rods. Damaged areas are cut or ground out, then welded back up to their original or an improved profile. Old turbine runners may have a significant amount of stainless steel added this way by the end of their lifetime. Elaborate welding procedures may be used to achieve the highest quality repairs.

Other elements requiring inspection and repair during overhauls include bearings, packing box and shaft sleeves, servomotors, cooling systems for the bearings and generator coils, seal rings, wicket gate linkage elements and all surfaces.

Environmental Impact

Water turbines are generally considered a clean power producer, as the turbine causes essentially no change to the water. They use a renewable energy source and are designed to operate for decades. They produce significant amounts of the world's electrical supply.

Historically there have also been negative consequences, mostly associated with the dams normally required for power production. Dams alter the natural ecology of rivers, potentially killing fish, stopping migrations, and disrupting peoples' livelihoods. For example, American Indian tribes in the Pacific Northwest had livelihoods built around salmon fishing, but aggressive dam-building destroyed their way of life. Dams also cause less obvious, but potentially serious consequences, including increased evaporation of water (especially in arid regions), buildup of silt behind the dam, and changes to water temperature and flow patterns. In the United States, it is now illegal to block the migration of fish, for example the white sturgeon in North America, so fish ladders must be provided by dam builders.

Water Wheel

An overshot waterwheel standing 42 ft (13 m) high powers the Old Mill at Berry College in Rome, Georgia, USA.

A water wheel is a machine for converting the energy of free-flowing or falling water into useful forms of power, often in a watermill. A water wheel consists of a large wooden or metal wheel, with a number of blades or buckets arranged on the outside rim forming the driving surface. Most commonly, the wheel is mounted vertically on a horizontal axle, but the tub or Norse wheel is mounted horizontally on a vertical shaft. Vertical wheels can transmit power either through the axle or via a ring gear and typically drive belts or gears; horizontal wheels usually directly drive their load.

Water wheel powering a mine hoist in De re metallica (1566).

Water wheels were still in commercial use well into the 20th century, but they are no longer in common use. Prior uses of water wheels include milling flour in gristmills and grinding wood into pulp for papermaking, but other uses include hammering wrought iron, machining, ore crushing and pounding fiber for use in the manufacture of cloth.

Some water wheels are fed by water from a mill pond, which is formed when a flowing stream is dammed. A channel for the water flowing to or from a water wheel is called a mill race (also spelled millrace) or simply a "race", and is customarily divided into sections. The race bringing water from the mill pond to the water wheel is a headrace; the one carrying water after it has left the wheel is commonly referred to as a tailrace.

John Smeaton's scientific investigation of the water wheel led to significant increases in efficiency in the mid to late 18th century and supplying much needed power for the Industrial Revolution.

Water wheels began being displaced by the smaller, less expensive and more efficient turbine, developed by Benoît Fourneyron, beginning with his first model in 1827. Turbines are capable of handling high *heads*, or elevations, that exceed the capability of practical-sized waterwheels.

The main difficulty of water wheels is their dependence on flowing water, which limits where they can be located. Modern hydroelectric dams can be viewed as the descendants of the water wheel, as they too take advantage of the movement of water downhill.

History

The two main functions of water wheels were historically water-lifting for irrigation purposes and as a power source. In terms of power source, water wheels can be turned either by human or animal force or by the water current itself. Water wheels come in two basic designs, either equipped with a vertical or a horizontal axle. The latter type can be subdivided, depending on where the water hits the wheel paddles, into overshot, breastshot and undershot wheels.

Greco-Roman World

Engineers of the Hellenistic era Mediterranean region are credited with the development of the water wheel. Mediterranean engineers of the Hellenistic and Roman periods were also the first to use it for both irrigation and as a power source. The technological breakthrough occurred in the technically advanced and scientifically minded Hellenistic period between the 3rd and 1st centuries BCE. This is seen as an evolution of the paddle-driven water-lifting wheels that had appeared in ancient Egypt by the 4th century BCE. According to John Peter Oleson, both the compartmented wheel and the hydraulic Noria appeared in Egypt by the 4th century BCE, with the Sakia being invented there a century later. This is supported by archeological finds at Faiyum, where the oldest archeological evidence of a water-wheel has been found, in the form of a Sakia dating back to the 3rd century BCE. A papyrus dating to the 2nd century BCE also found in Faiyum mentions a water wheel used for irrigation, a 2nd-century BC fresco found at Alexandria depicts a compartmented Sakia, and the writings of Callixenus of Rhodes mention the use of a Sakia in Ptolemaic Egypt during the reign of Ptolemy IV in the late 3rd century BC.

Drainage Wheels

The Romans used water wheels extensively in mining projects. Several such devices were described by Vitruvius. The one found during modern mining at the copper mines at Rio Tinto in Spain involved 16 such wheels stacked above one another so as to lift water about 80 feet (24 m) from the mine sump. Part of a similar wheel dated to about 90 CE, was found in the 1930s, at Dolaucothi, a Roman gold mine in south Wales.

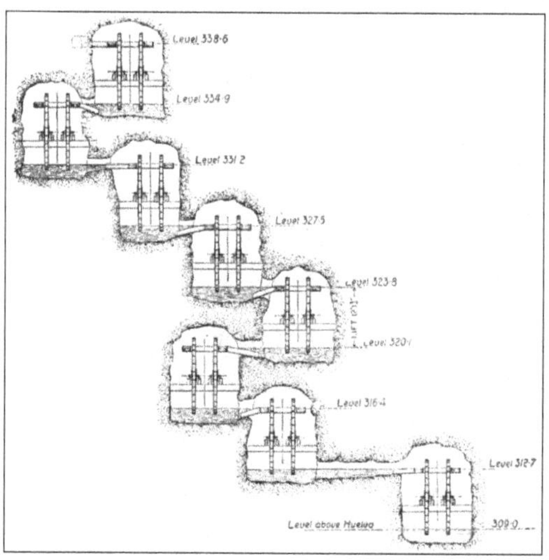

Sequence of wheels found in Rio Tinto mines, southwestern Spain.

Water Mills

Taking indirect evidence into account from the work of the Greek technician Apollonius of Perge, the British historian of technology M.J.T. Lewis dates the appearance of the vertical-axle watermill to the early 3rd century BCE, and the horizontal-axle watermill to around 240 BC, with Byzantium and Alexandria as the assigned places of invention. A watermill is reported by the Greek geographer Strabon (ca. 64 BCE–CE 24) to have existed sometime before 71 BCE in the palace of the Pontian king Mithradates VI Eupator, but its exact construction cannot be gleaned from the text (XII, 3, 30 C 556).

Reconstruction of Vitruvius' undershot-wheeled watermill.

The first clear description of a geared watermill is from the 1st-century BC Roman architect Vitruvius, who tells of the sakia gearing system as being applied to a watermill. Vitruvius's account is particularly valuable in that it shows how the watermill came about, namely by the combination of the separate Greek inventions of the toothed gear and the water wheel into one effective mechanical system for harnessing water power. Vitruvius's water wheel is described as being immersed with its lower end in the watercourse so that its paddles could be driven by the velocity of the running water (X, 5.2).

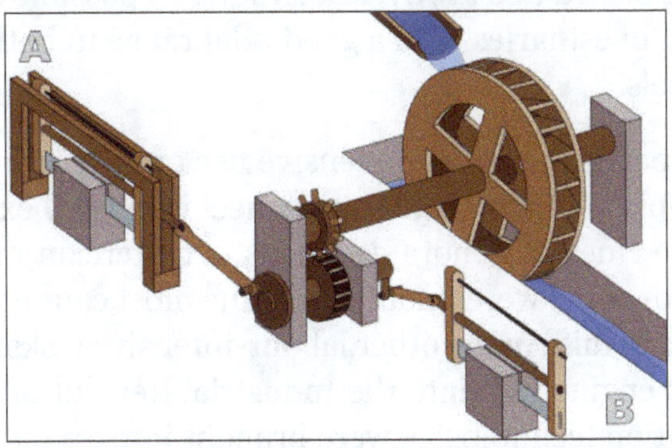

Schematic of the Roman Hierapolis sawmill, Asia Minor, powered by a breastshot wheel.

About the same time, the overshot wheel appears for the first time in a poem by Antipater of Thessalonica, which praises it as a labour-saving device (IX, 418.4–6). The motif is also taken up by Lucretius (ca. 99-55 BC) who likens the rotation of the water wheel to the motion of the stars on the firmament (V 516). The third horizontal-axled type, the breastshot water wheel, comes into archaeological evidence by the late-2nd-century AD context in central Gaul. Most excavated Roman watermills were equipped with one of these wheels which, although more complex to construct, were much more efficient than the vertical-axle water wheel. In the 2nd century AD, Barbegal watermill complex a series of sixteen overshot wheels was fed by an artificial aqueduct, a proto-industrial grain factory which has been referred to as "the greatest known concentration of mechanical power in the ancient world".

In Roman North Africa, several installations from around 300 AD were found where vertical-axle water wheels fitted with angled blades were installed at the bottom of a water-filled, circular shaft. The water from the mill-race which entered the pit tangentially created a swirling water column that made the fully submerged wheel act like true water turbines, the earliest known to date.

Navigation

Apart from its use in milling and water-raising, ancient engineers applied the paddled water wheel for automatons and in navigation. Vitruvius (X 9.5-7) describes multi-geared paddle wheels working as a ship odometer, the earliest of its kind. The first mention of paddle wheels as a means of propulsion comes from the 4th–5th-century military treatise *De Rebus Bellicis* (chapter XVII), where the anonymous Roman author describes an ox-driven paddle-wheel warship.

Early Medieval Europe

Ancient water-wheel technology continued unabated in the early medieval period where the appearance of new documentary genres such as legal codes, monastic charters, but also hagiography was accompanied with a sharp increase in references to watermills and wheels.

The earliest vertical-wheel in a tide mill is from 6th-century Killoteran near Waterford, Ireland, while the first known horizontal-wheel in such a type of mill is from the Irish Little Island (c. 630). As for the use in a common Norse or Greek mill, the oldest known horizontal-wheels were excavated in the Irish Ballykilleen, dating to c. 636.

The earliest excavated water wheel driven by tidal power was the Nendrum Monastery mill in Northern Ireland which has been dated at 787A.D. although a possible earlier mill dates to 619A.D. Tide mills became common in estuaries with a good tidal range in both Europe and America generally using undershot wheels.

Cistercian monasteries, in particular, made extensive use of water wheels to power watermills of many kinds. An early example of a very large water wheel is the still extant wheel at the early 13th century Real Monasterio de Nuestra Senora de Rueda, a Cistercian monastery in the Aragon region of Spain. Grist mills (for corn) were undoubtedly the most common, but there were also sawmills, fulling mills and mills to fulfil many other labour-intensive tasks. The water wheel remained competitive with the steam engine well into the Industrial Revolution. At around the 8th to 10th century, a number of irrigation technologies were brought into Spain and thus introduced to Europe. One of those technologies is the Noria, which is basically a wheel fitted with buckets on the peripherals for lifting water. It is similar to the undershot water wheel mentioned later in this article. It allowed peasants to power watermills more efficiently. According to Thomas Glick's book,

Irrigation and Society in Medieval Valencia, the Noria probably originated from somewhere in Persia. It has been used for centuries before the technology was brought into Spain by Arabs who had adopted it from the Romans. Thus the distribution of the Noria in the Iberian peninsula "conforms to the area of stabilized Islamic settlement". This technology has a profound effect on the life of peasants. The Noria is relatively cheap to build. Thus it allowed peasants to cultivate land more efficiently in Europe. Together with the Spaniards, the technology then spread to North Africa and later to the New World in Mexico and South America following Spanish expansion.

Water wheel powering a small village mill at the Museum of Folk Architecture and Life, Uzhhorod, Ukraine.

Domesday Inventory of English Mills ca. 1086

The assembly convened by William of Normandy, commonly referred to as the "Domesday" or Doomsday survey, took an inventory of all potentially taxable property in England, which included over six thousand mills spread across three thousand different locations.

Locations

The type of water wheel selected was dependent upon the location. Generally if only small volumes of water and high waterfalls were available a millwright would choose to use an overshot wheel. The decision was influenced by the fact that the buckets could catch and use even a small volume of water. For large volumes of water with small waterfalls the undershot wheel would have been used, since it was more adapted to such conditions and cheaper to construct. So long as these water supplies were abundant the question of efficiency remained irrelevant. By the 18th century with increased demand for power coupled with limited water locales, an emphasis was made on efficiency scheme.

Economic Influence

By the 11th century there were parts of Europe where the exploitation of water was commonplace. The water wheel is understood to have actively shaped and forever changed the outlook of Westerners. Europe began to transit from human and animal muscle labor towards mechanical labor with the advent of the water wheel. Medievalist Lynn White Jr. contended that the spread of inanimate power sources was eloquent testimony to the emergence of the West of a new attitude toward, power, work, nature, and above all else technology.

Harnessing water-power enabled gains in agricultural productivity, food surpluses and the large scale urbanization starting in the 11th century. The usefulness of water power motivated European experiments with other power sources, such as wind and tidal mills. Waterwheels influenced the construction of cities, more specifically canals. The techniques that developed during this early period such as stream jamming and the building of canals, put Europe on a hydraulically focused path, for instance water supply and irrigation technology was combined to modify supply power of the wheel. Illustrating the extent to which there was a great degree of technological innovation that met the growing needs of the feudal state.

Applications of The Water Wheel in Medieval Europe

The water mill was used for grinding grain, producing flour for bread, malt for beer, or coarse meal for porridge. Hammermills used the wheel to operate hammers. One type was fulling mill, which was used for cloth making. The trip hammer was also used for making wrought iron and for working iron into useful shapes, an activity that was otherwise labour-intensive. The water wheel was also used in papermaking, beating material to a pulp. In the 13th century water mills used for hammering throughout Europe improved the productivity of early steel manufacturing. Along with the mastery of gunpowder, waterpower provided European countries worldwide military leadership from the 15th century.

Ore stamp mill (behind worker taking ore form chute). From Georg Agricola's De re metallica (1556).

Importance to 17th- and 18th-century Europe (Scientific Influence)

Millwrights distinguished between the two forces, impulse and weight, at work in water wheels long before 18th-century Europe. Fitzherbert, a 16th-century agricultural writer, wrote "druieth the wheel as well as with the weight of the water as with strengthe [impulse]." Leonardo da Vinci also discussed water power, noting "the blow [of the water] is not weight, but excites a power of weight, almost equal to its own power." However, even realisation of the two forces, weight and impulse, confusion remained over the advantages and disadvantages of the two, and there was no clear understanding of the superior efficiency of weight. Prior to 1750 it was unsure as to which force was dominant and was widely understood that both forces were operating with equal inspiration amongst one another. The waterwheel, sparked questions of the laws of nature, specifically the laws of force. Evangelista Torricelli's work on water wheels used an analysis of Galileo's work on falling bodies, that the velocity of a water sprouting from an orifice under its head was exactly equivalent to the velocity a drop of water acquired in falling freely from the same height.

Industrial European Usage

The most powerful water wheel built in the United Kingdom was the 100 hp Quarry Bank Mill water wheel near Manchester. A high breastshot design, it was retired in 1904 and replaced with several turbines. It has now been restored and is a museum open to the public.

The biggest working water wheel in mainland Britain has a diameter of 15.4 m and was built by the De Winton company of Caernarfon. It is located within the Dinorwic workshops of the National Slate Museum in Llanberis, North Wales.

The largest working water wheel in the world is the Laxey Wheel (also known as *Lady Isabella*) in the village of Laxey, Isle of Man. It is 72 feet 6 inches (22.10 m) in diameter and 6 feet (1.83 m) wide and is maintained by Manx National Heritage.

Lady Isabella Wheel, Laxey, Isle of Man, used to drive mine pumps.

Development of water turbines during the Industrial Revolution led to decreased popularity of water wheels. The main advantage of turbines is that its ability to harness head is much greater than the diameter of the turbine, whereas a water wheel cannot effectively harness head greater than its diameter. The migration from water wheels to modern turbines took about one hundred years.

A mid-19th-century water wheel at Cromford in England used for grinding locally mined barytes.

China

Chinese water wheels almost certainly have a separate origin, as early ones there were invariably horizontal water wheels. By at least the 1st century AD, the Chinese of the Eastern Han Dynasty were using water wheels to crush grain in mills and to power the piston-bellows in forging iron ore into cast iron.

In the text known as the *Xin Lun* written by Huan Tan about 20 AD (during the usurpation of Wang Mang), it states that the legendary mythological king known as Fu Xi was the one responsible for the pestle and mortar, which evolved into the tilt-hammer and then trip hammer device. Although the author speaks of the mythological Fu Xi, a passage of his writing gives hint that the water wheel was in widespread use by the 1st century AD in China.

Fu Hsi invented the pestle and mortar, which is so useful, and later on it was cleverly improved in such a way that the whole weight of the body could be used for treading on the tilt-hammer (*tui*), thus increasing the efficiency ten times. Afterwards the power of animals—donkeys, mules, oxen, and horses—was applied by means of machinery, and water-power too used for pounding, so that the benefit was increased a hundredfold.

In the year 31 AD, the engineer and Prefect of Nanyang, Du Shi (d. 38), applied a complex use of the water wheel and machinery to power the bellows of the blast furnace to create cast iron. Du Shi is mentioned briefly in the *Book of Later Han* (*Hou Han Shu*) as follows (in Wade-Giles spelling):

In the seventh year of the Chien-Wu reign period (31 AD) Tu Shih was posted to be Prefect of

Nanyang. He was a generous man and his policies were peaceful; he destroyed evil-doers and established the dignity (of his office). Good at planning, he loved the common people and wished to save their labor. He invented a water-power reciprocator (*shui phai*) for the casting of (iron) agricultural implements. Those who smelted and cast already had the push-bellows to blow up their charcoal fires, and now they were instructed to use the rushing of the water (*chi shui*) to operate it ... Thus the people got great benefit for little labor. They found the 'water(-powered) bellows' convenient and adopted it widely.

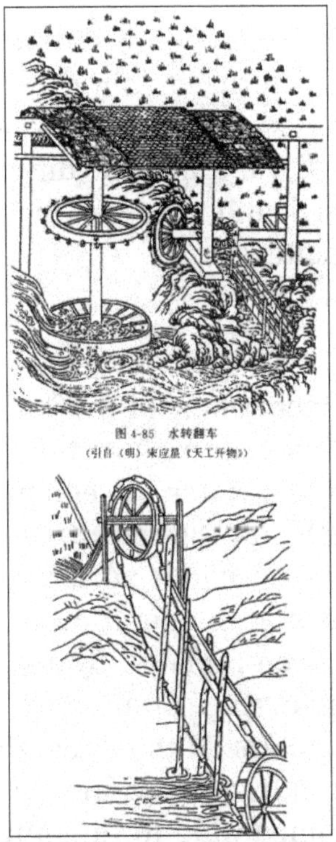

Two types of hydraulic-powered chain pumps from the Tiangong Kaiwu of 1637, written by the Ming Dynasty encyclopedist, Song Yingxing (1587–1666).

Water wheels in China found practical uses such as this, as well as extraordinary use. The Chinese inventor Zhang Heng (78–139) was the first in history to apply motive power in rotating the astronomical instrument of an armillary sphere, by use of a water wheel. The mechanical engineer Ma Jun (c. 200–265) from Cao Wei once used a water wheel to power and operate a large mechanical puppet theater for the Emperor Ming of Wei.

India

The early history of the watermill in India is obscure. Ancient Indian texts dating back to the 4th century BC refer to the term *cakkavattaka* (turning wheel), which commentaries explain as *arahatta-ghati-yanta* (machine with wheel-pots attached). On this basis, Joseph Needham suggested that the machine was a noria. Terry S. Reynolds, however, argues that the "term used in Indian texts is ambiguous and does not clearly indicate a water-powered device." Thorkild Schiøler argued that it is "more likely that these passages refer to some type of tread- or hand-operated water-lifting device, instead of a water-powered water-lifting wheel."

According to Greek historical tradition, India received water-mills from the Roman Empire in the early 4th century AD when a certain Metrodoros introduced "water-mills and baths, unknown among them [the Brahmans] till then". Irrigation water for crops was provided by using water raising wheels, some driven by the force of the current in the river from which the water was being raised. This kind of water raising device was used in ancient India, predating, according to Pacey, its use in the later Roman Empire or China, even though the first literary, archaeological and pictorial evidence of the water wheel appeared in the Hellenistic world.

Around 1150, the astronomer Bhaskara Achārya observed water-raising wheels and imagined such a wheel lifting enough water to replenish the stream driving it, effectively, a perpetual motion machine. The construction of water works and aspects of water technology in India is described in Arabic and Persian works. During medieval times, the diffusion of Indian and Persian irrigation technologies gave rise to an advanced irrigation system which brought about economic growth and also helped in the growth of material culture.

Islamic World

Arab engineers took over the water technology of the hydraulic societies of the ancient Near East; they adopted the Greek water wheel as early as the 7th century, excavation of a canal in the Basra region discovered remains of a water wheel dating from this period. Hama in Syria still preserves some of its large wheels, on the river Orontes, although they are no longer in use. One of the largest had a diameter of about 20 metres and its rim was divided into 120 compartments. Another wheel that is still in operation is found at Murcia in Spain, La Nora, and although the original wheel has been replaced by a steel one, the Moorish system during al-Andalus is otherwise virtually unchanged. Some medieval Islamic compartmented water wheels could lift water as high as 30 meters. Muhammad ibn Zakariya al-Razi's *Kitab al-Hawi* in the 10th century described a noria in Iraq that could lift as much as 153,000 litres per hour, or 2550 litres per minute. This is comparable to the output of modern norias in East Asia, which can lift up to 288,000 litres per hour, or 4800 litres per minute.

The norias of Hama on the Orontes River.

The industrial uses of watermills in the Islamic world date back to the 7th century, while horizontal-wheeled and vertical-wheeled water mills were both in widespread use by the 9th century. A variety of industrial watermills were used in the Islamic world, including gristmills, hullers, sawmills, shipmills, stamp mills, steel mills, sugar mills, and tide mills. By the 11th century, every province throughout the Islamic world had these industrial watermills in operation, from al-Andalus and North Africa to the Middle East and Central Asia. Muslim and Christian engineers also used crankshafts and water turbines, gears in watermills and water-raising machines, and dams as a source of water, used to provide additional power to watermills and water-raising machines. Fulling mills and steel mills may have spread from Islamic Spain to Christian Spain in the 12th century. Industrial water mills were also employed in large factory complexes built in al-Andalus between the 11th and 13th centuries.

Water wheel in Djambi, Sumatra, c. 1918.

The engineers of the Islamic world developed several solutions to achieve the maximum output from a water wheel. One solution was to mount them to piers of bridges to take advantage of the increased flow. Another solution was the shipmill, a type of water mill powered by water wheels mounted on the sides of ships moored in midstream. This technique was employed along the Tigris and Euphrates rivers in 10th-century Iraq, where large shipmills made of teak and iron could produce 10 tons of flour from corn every day for the granary in Baghdad. The flywheel mechanism, which is used to smooth out the delivery of power from a driving device to a driven machine, was invented by Ibn Bassal (fl. 1038-1075) of Al-Andalus; he pioneered the use of the flywheel in the saqiya (chain pump) and noria. The engineers Al-Jazari in the 13th century and Taqi al-Din in the 16th century described many inventive water-raising machines in their technological treatises. They also employed water wheels to power a variety of devices, including various water clocks and automata.

Types

Most water wheels in the United Kingdom and the United States are (or were) vertical wheels rotating about a horizontal axle, but in the Scottish highlands and parts of southern Europe mills often had a horizontal wheel (with a vertical axle). Water wheels are classified by the way in which water is applied to the wheel, relative to the wheel's axle. Overshot and pitchback water wheels are suitable where there is a small stream with a height difference of more than 2 meters, often in association with a small reservoir. Breastshot and undershot wheels can be used on rivers or high volume flows with large reservoirs.

Horizontal Wheel

Commonly called a tub wheel or Norse mill, the horizontal wheel is essentially a very primitive and inefficient form of the modern turbine. It is usually mounted inside a mill building below the working floor. A jet of water is directed on to the paddles of the water wheel, causing them to turn; water exits beneath the wheel, generally through the center. This is a simple system, usually used without gearing so that the vertical axle of the water wheel becomes the drive spindle of the mill.

Undershot Wheel

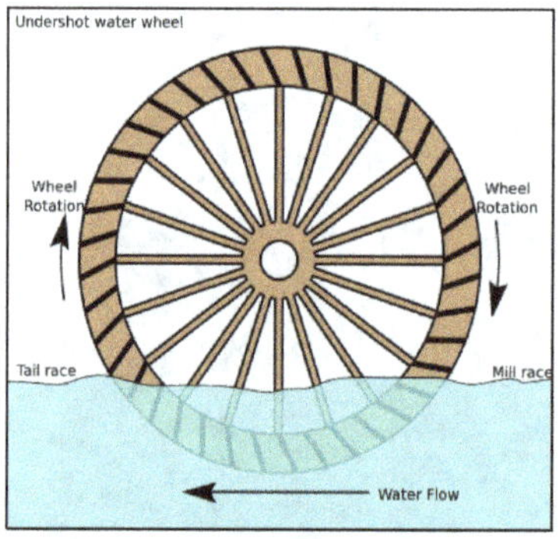

Undershot water wheel.

An undershot wheel (also called a *stream wheel*) is a vertically mounted water wheel that is rotated by water striking paddles or blades at the bottom of the wheel. The name *undershot* comes from this striking at the bottom of the wheel. This type of water wheel is the oldest type of wheel.

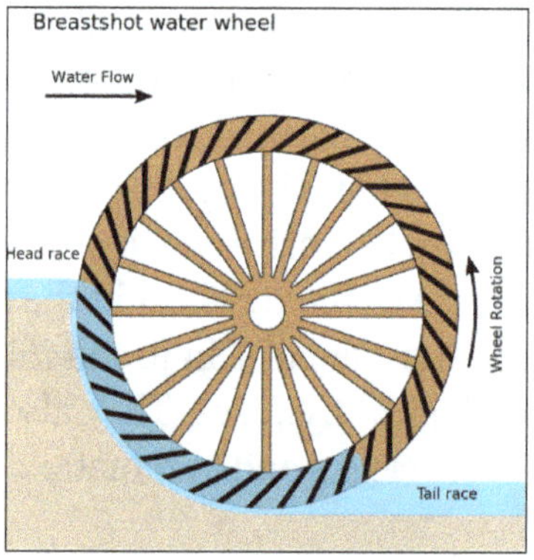

Breastshot water wheel.

It is also regarded as the least efficient type, although subtypes of this water wheel (e.g. the Poncelet wheel, Sagebien wheel and Zuppinger wheel) allow somewhat greater efficiencies than the traditional undershot wheels. The advantages of undershot wheels are that they are somewhat cheaper and simpler to build, and have less of an environmental impact—as they do not constitute a major change of the river. Their disadvantages are—as mentioned before—less efficiency, which means that they generate less power and can only be used where the flow rate is sufficient to provide torque.

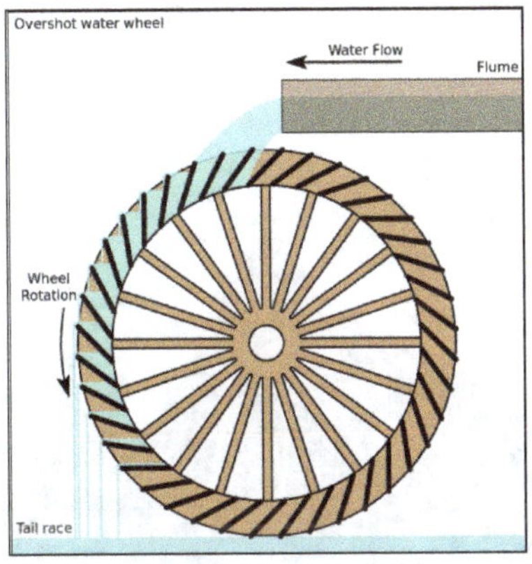

Overshot water wheel.

Undershot wheels gain no advantage from head. They are most suited to shallow streams in flat country.

Undershot wheels are also well suited to installation on floating platforms. The earliest were probably constructed by the Byzantine general Belisarius during the siege of Rome in 537. Later they were sometimes mounted immediately downstream from bridges where the flow restriction of arched bridge piers increased the speed of the current.

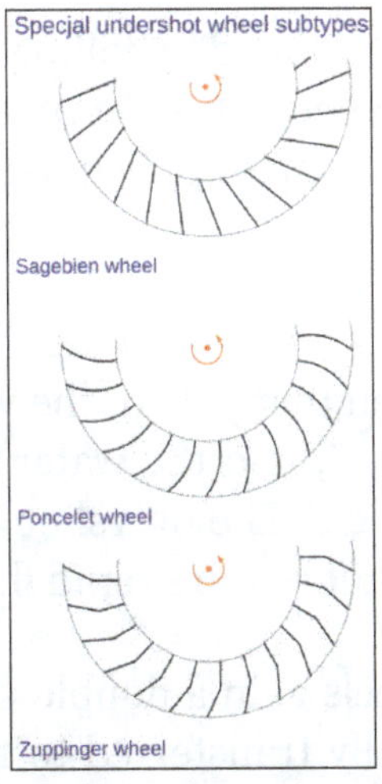

Sabegien, Poncelet and Zuppinger water wheel.

Breastshot Wheel

A vertically mounted water wheel that is rotated by falling water striking buckets near the center of the wheel's edge, or just above it, is said to be *breastshot*. Breastshot wheels are the most common type in the United States of America and are said to have powered the American industrial revolution.

Breastshot wheels are less efficient than overshot wheels, are more efficient than undershot wheels, and are not backshot. The individual blades of a breastshot wheel are actually buckets, as are those of most overshot wheels, and not simple paddles like those of most undershot wheels. A breastshot wheel requires a good trash rack and typically has a masonry "apron" closely conforming to the wheel face, which helps contain the water in the buckets as they progress downwards. Breastshot wheels are preferred for steady, high-volume flows such as are found on the fall line of the North American East Coast.

The Anderson Mill of Texas is undershot, backshot, and overshot using two sources of water. This allows the speed of the wheel to be controlled.

Overshot Wheel

A vertically mounted water wheel that is rotated by falling water striking paddles, blades or buckets near the top of the wheel is said to be *overshot*. In true overshot wheels the water passes over the top of the wheel, but the term is sometimes applied to backshot or pitchback wheels where the water goes down behind the water wheel.

A typical overshot wheel has the water channeled to the wheel at the top and slightly beyond the axle. The water collects in the buckets on that side of the wheel, making it heavier than the other "empty" side. The weight turns the wheel, and the water flows out into the tail-water when the wheel rotates enough to invert the buckets. The overshot design can use all of the water flow for power (unless there is a leak) and does not require rapid flow.

Unlike undershot wheels, overshot wheels gain a double advantage from gravity. Not only is the momentum of the flowing water partially transferred to the wheel, the weight of the water descending in the wheel's buckets also imparts additional energy. The mechanical power derived from an overshot wheel is determined by the wheel's physical size and the available head, so they are ideally suited to hilly or mountainous country. On average, the undershot wheel uses 22 percent of the energy in the flow of water, while an overshot wheel uses 63 percent, as calculated by English civil engineer John Smeaton in the 18th century.

Overshot wheels demand exact engineering and significant head, which usually means significant investment in constructing a dam, millpond and waterways. Sometimes the final approach of the water to the wheel is along a lengthy flume or penstock.

Reversible Wheel

A special type of overshot wheel is the reversible water wheel. This has two sets of blades or buckets running in opposite directions, so that it can turn in either direction depending on which side the water is directed. Reversible wheels were used in mining industry in order to power various means of ore conveyance. By changing the direction of the wheel, barrels or baskets of ore could be lifted up or lowered down a shaft. As a rule there was also a cable drum or a chain basket (German: *Kettenkorb*) on the axle of the wheel. It was also essential that the wheel had braking equipment in order to be able to stop the wheel (known as a braking wheel). The oldest known drawing of a reversible water wheel was by Georgius Agricola and dates to 1556.

Replica of a reversible wheel with a 9.5 m diameter in Clausthal-Zellerfeld.

Backshot Wheel

A backshot wheel (also called *pitchback*) is a variety of overshot wheel where the water is introduced just behind the summit of the wheel. It combines the advantages from breastshot and overshot systems, since the full amount of the potential energy released by the falling water is harnessed as the water descends the back of the wheel (as in overshot wheel) while it also gains power from the water's current past the bottom of the wheel (as in breastshot wheel).

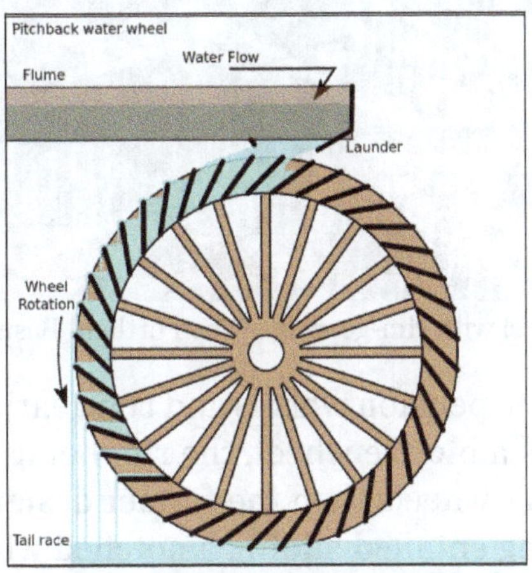

Pitchback or "backshot" water wheel.

A backshot wheel continues to function until the water in the wheel pit rises well above the height of the axle, when any other overshot wheel will be stopped or even destroyed. This makes the technique particularly suitable for streams that experience extreme seasonal variations in flow, and reduces the need for complex sluice and tail race configurations.

Backshot wheel at New Lanark World Heritage Site, Scotland.

The direction of rotation of a backshot wheel is the same as that of a breastshot wheel at the same location so it can easily replace one, without causing the directional gearing in the mill to be changed. This would increase the power available while only requiring a change to be made to the water level in the top pond, which in some cases is economically viable.

Suspension Wheels and Rim-gears

The suspension wheel with rim-gearing at the Portland Basin Canal Warehouse.

Two early improvements were suspension wheels and rim gearing. Suspension wheels are constructed in the same manner as a bicycle wheel, the rim being supported under tension from the hub- this led to larger lighter wheels than the former design where the heavy spokes were under compression. Rim-gearing entailed adding a notched wheel to the rim or shroud of the wheel. A stub gear engaged the rim-gear and took the power into the mill using an independent line shaft. This removed the rotative stress from the axle which could thus be lighter, and also allowed more flexibility in the location of the power train. The shaft rotation was geared up from that of the wheel which led to less power loss. An example of this design pioneered by Thomas Hewes and refined by William Fairburn can be seen at the 1849 restored wheel at the Portland Basin Canal Warehouse.

Efficiency

Overshot (and particularly backshot) wheels are the most efficient type; a backshot steel wheel can be more efficient (about 60%) than all but the most advanced and well-constructed turbines. In some situations an overshot wheel is preferable to a turbine.

The development of the hydraulic turbine wheels with their improved efficiency (>67%) opened up an alternative path for the installation of water wheels in existing mills, or redevelopment of abandoned mills.

Power Calculations

In an undershot wheel or a run of the river wheel the power is dependant to the kinetic energy of the river. Approximate power can be calculated.

Power in Watts= $100 \times A \times V^3 \times C$

A = Area of paddles in the water (square meters).

V = Velocity of the stream in meters per second.

C = Efficiency Constant (assume 1 for a water to wire efficiency of 20%).

Rotational speed of the wheel = $9 \times V / D$ rpm.

D = diameter in metres.

The great water wheel in the Welsh National Slate Museum.

For a breast shot or over shot wheel both potential energy and kinetic energy must be considered. This takes the form of the weight of water in the buckets and the vertical distance travelled. A rule of thumb formula is

Power in Watts = $4 \times Q \times H \times C$

Q = Weight of water (volume per sec x capacity of the buckets).

V = Velocity of the stream in meters per second.

H = Head, or height difference of water between the lip of the flume (head race) and the tailrace.

C = Efficiency Constant.

The optimal rotational speed of a breast shot or overshot wheel is approximately:

Rotational speed of the wheel= $21/\sqrt{D}$

D = diameter of the wheel in metres.

Hydraulic Wheel

A recent development of the breastshot wheel is a hydraulic wheel which effectively incorporates automatic regulation systems. The Aqualienne is one example. It generates between 37 kW and 200 kW of electricity from a 20m^3 waterflow with a head of 1 to 3.5m. It is designed to produce electricity at the sites of former watermills.

Hydraulic Wheel Part Reaction Turbine

A parallel development is the hydraulic wheel/part reaction turbine that also incorporates a weir into the centre of the wheel but uses blades angled to the water flow. The WICON-Stem Pressure Machine (SPM) exploits this flow. Estimated efficiency 67%.

The University of Southampton School of Civil Engineering and the Environment in the UK has investigated both types of Hydraulic wheel machines and has estimated their hydraulic efficiency and suggested improvements, i.e. The Rotary Hydraulic Pressure Machine. (Estimated maximum efficiency 85%).

These type of water wheels have high efficiency at part loads / variable flows and can operate at very low heads, < 1 metre. Combined with direct drive Axial Flux Permanent Magnet Alternators and power electronics they offer a viable alternative for low head hydroelectric power generation.

Water-lifting

In water-raising devices rotary motion is typically more efficient than machines based on oscillating motion.

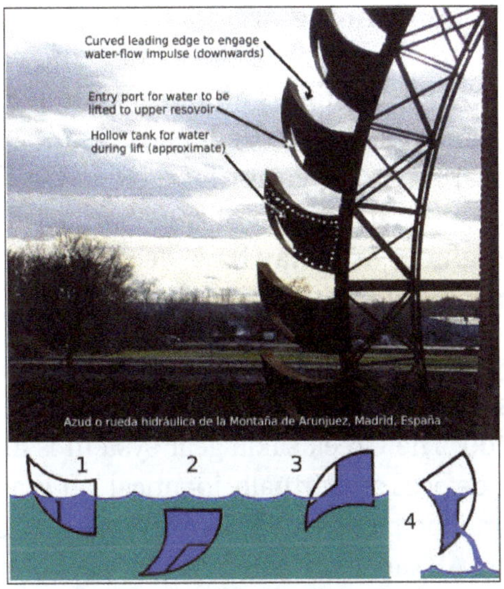

Detail of azud at Aranjuez, Spain.

The compartmented water wheel comes in two basic forms, the wheel with compartmented body (Latin *tympanum*) and the wheel with compartmented rim or a rim with separate, attached containers. The wheels could be either turned by the flow of water, men treading on its outside or by animals by means of a sakia gear. While the tympanum had a large discharge capacity, it could lift the water only to less than the height of its own radius and required a large torque for rotating. These constructional deficiencies were overcome by the wheel with a compartmented rim which was a less heavy design with a higher lift.

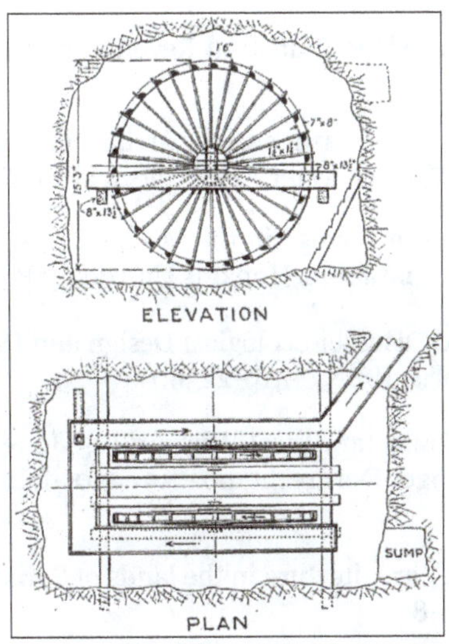

Drainage wheel from Rio Tinto mines.

Ptolemaic Egypt

The earliest literary reference to a water-driven, compartmented wheel appears in the technical treatise *Pneumatica* (chap. 61) of the Greek engineer Philo of Byzantium (ca. 280–220 BC). In his *Parasceuastica* (91.43–44), Philo advises the use of such wheels for submerging siege mines as a defensive measure against enemy sapping. Compartmented wheels appear to have been the means of choice for draining dry docks in Alexandria under the reign of Ptolemy IV (221–205 BC). Several Greek papyri of the 3rd to 2nd century BC mention the use of these wheels, but don't give further details. The non-existence of the device in the Ancient Near East before Alexander's conquest can be deduced from its pronounced absence from the otherwise rich oriental iconography on irrigation practices. Unlike other water-lifting devices and pumps of the period though, the invention of the compartmented wheel cannot be traced to any particular Hellenistic engineer and may have been made in the late 4th century BC in a rural context away from the metropolis of Alexandria.

The earliest depiction of a compartmented wheel is from a tomb painting in Ptolemaic Egypt which dates to the 2nd century BC. It shows a pair of yoked oxen driving the wheel via a sakia gear, which is here for the first time attested, too. The Greek sakia gear system is already shown fully developed to the point that "modern Egyptian devices are virtually identical". It is assumed that the scientists of the Museum of Alexandria, at the time the most active Greek research center, may have been involved in its invention. An episode from the Alexandrian War in 48 BC tells of how Caesar's enemies employed geared water wheels to pour sea water from elevated places on the position of the trapped Romans.

Around 300 AD, the noria was finally introduced when the wooden compartments were replaced with inexpensive ceramic pots that were tied to the outside of an open-framed wheel.

References

- Al-Hassani, Salim. "800 Years Later: In Memory of Al-Jazari, A Genius Mechanical Engineer". Muslim Heritage. The Foundation for Science, Technology, and Civilisation. Retrieved 30, March 2020
- Kreis, Steven (2001). "The Origins of the Industrial Revolution in England". The history guide. Retrieved 08, August 2020
- *Nevell, Mike; Walker (2001). Portland Basin and the archaeology of the Canal Warehouse. Tameside Metropolitan Borough with University of Manchester Archaeological Unit. ISBN 1-871324-25-4
- Thomson, Ross (2009). Structures of Change in the Mechanical Age: Technological Invention in the United States 1790-1865. Baltimore, MD: The Johns Hopkins University Press. p. 34. ISBN 978-0-8018-9141-0
- William R. Jobin, 1999. Dams and Disease: Ecological Design and Health Impacts of Large Dams, Canals, and Irrigation Systems, Taylor & Francis, ISBN 0-419-22360-6
- Hurford, Clive.; Schneider, M; Cowx, Ian; West, Richard (25 June 1997), "21", Conservation Monitoring in Freshwater Habitats, Berlin: Springer Dordrecht, pp. 219–230, doi:10.1007/978-1-4020-9278-7, ISBN 978-1-4020-9277-0, ISSN 0343-6993
- Adriana de Miranda (2007). Water architecture in the lands of Syria: the water-wheels. L'Erma di Bretschneider. pp. 38–9. ISBN 88-8265-433-8
- Bent Sørensen (2004). Renewable Energy: Its Physics, Engineering, Use, Environmental Impacts, Economy, and Planning Aspects. Academic Press. pp. 556–. ISBN 978-0-12-656153-1

5

Biomass as a Source of Energy

Biomass energy refers to the generation of energy by organic materials. Biomass is any plant material used as a source of fuel. Biofuel and biogas are two renewable energy sources derived from organic wastes like food scraps, animal wastes, feedstock, etc. Biomass energy and its features are explained in detail in this chapter.

Biomass Energy

In the past few years, there have been significant improvements in renewable energy technologies along with declines in cost. The growing concern for the environment and sustainable development, have led to worldwide interest in renewable energies and bio- energy in particular. Biomass can be converted into modern energy forms such as liquid and gaseous fuels, electricity, and process heat to provide energy services needed by rural and urban populations and also by industry. This paper explains the different ways of extracting energy from biomass and a comparison is made among them. This paper also explains about the potentiality of biomass energy in India, analyses current situation compares bio-energy and other options for promoting development, brings out the advantages over the other renewables putting forth the drawbacks to be overcome to make it still more successful. This paper analyses current situation compares bio-energy and other options for promoting development, explore the potential for bio-energy.

Introduction

In past 10 years or so, considerable practical experience has accumulated in India as well as in other developing and industrialized countries, on biomass energy production and conversion. India is pioneer among developing countries, with significant indigenous efforts in promoting renewable energy technologies. The importance of bio- energy as a modern fuel has been recognized. India has about 70,000 villages yet to be connected to the electricity grid. The supply of grid power to rural areas is characterized by

a) Low loads,

b) Power shortages,

c) Low reliability,

d) Low and fluctuating voltages,

e) High transmission and distribution costs and power losses.

Decentralized power generation based on renewables is an attractive option to meet the energy needs. The availability of biomass such as wood, cow-dung, leaf litter in rural areas is more. Hence a choice of biomass energy especially in rural areas is more reasonable but at the same time the technology is being developed to meet the large- scale requirements using biomass.

Biomass energy has played a key role in the time of Second World War when there was a fuel deficiency. Many vehicles, tractors and trucks used wood gasifies, which generate producer's gas, running an internal combustion Engine. One of the major advantages of biomass energy is that it can be used in different forms. For e.g., Gas generated from the biomass can be directly used for cooking or it can be used for running an internal combustion Engine for developing stationary shaft power or otherwise coupled to generator for generating electric power.

The subsequent sections explain about the different ways of extracting energy from biomass, explaining about technological and economic aspects followed by a case study. The issue of land availability for biomass (wood) production.

Motivation

There are several renewable and non-renewable energy options for power generation at the decentralized level. It is necessary to understand why biomass-based energy options should receive priority over other options and to discuss the advantages to local and global communities as well as the environment.

Biomass is renewable fuel used in nearly every corner of the developing countries as a source of heat, particularly in the domestic sector. Biomass energy includes energy from all plant matter (tree, shrub, and crop) and animal dung. Biomass, unlike other renewables, is a versatile source of energy, which can be converted to modern forms such as liquid and gaseous fuels, electricity, and process heat. For example, small-scale (5-100 KW), medium-scale (1-10 MW), and large scale (about 50 MW).

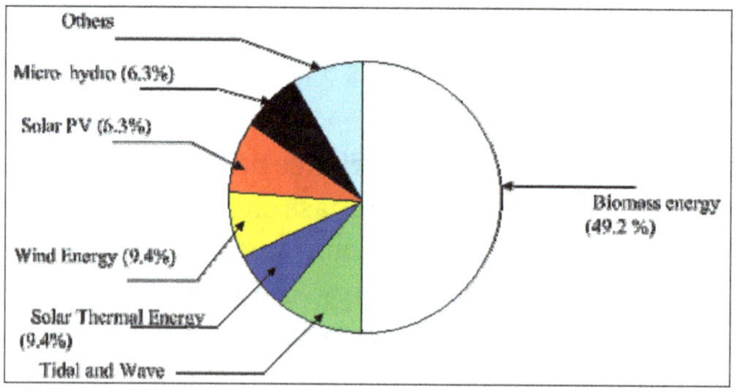

Expenditure on different renewable energy programmes for 1992-1997 indicating the share of bioenergy technologies for renewables in India.

It can be seen from the figure the importance of biomass energy in Indian context.

Advantages of biomass gasify energy over other renewable energy options:

1. Suitable in most locations.
2. Varying capacity can be installed; any capacity can be operated, even at lower loads; no seasonality.
3. Need for storage of energy is not required.

Advantages and drawbacks of biogas energy over other renewable energy options:

1. It can be used directly for cooking, or heating water from the abundantly available dung and dried plant leaves in rural areas.
2. Capacity determined by availability of dung. Not suitable for varying loads.
3. Not feasible to locate at all the locations.

Different Ways of Extracting Energy from Biomass

The different methods of biomass extraction can be broadly be classified as:

1. Anaerobic Digestion,
2. Gasification,
3. Liquefaction,

(a) Solid fuel combustion

The simplest and most common way of extracting energy from biomass is by direct combustion of solid matter. Majority of the developing countries especially in rural areas obtain the majority of their energy needs from the burning of wood, animal dung and other biomass. But burning can be inefficient. An open fireplace may let large amounts of heat escape, while a significant proportion of the fuel may not even get burnt.

(b) Gasification

Gasification is a process that exposes a solid fuel to high temperatures and limited oxygen, to produce a gaseous fuel. This is a mix of gases such as carbon monoxide, carbon dioxide, nitrogen, hydrogen and methane.

Gasification has several advantages over burning solid fuel. One is convenience – one of the resultant gases, methane, can be treated in a similar way as natural gas, and used for the same purposes.

Another advantage of gasification is that it produces a fuel that has had many impurities removed and will therefore cause fewer pollution problems when burnt. And, under suitable circumstances, it can produce synthesis gas, a mixture of carbon monoxide and hydrogen. This can be used to make almost any hydrocarbon (e.g., methane and methanol), which can then be substituted for fossil fuels. But hydrogen itself is a potential fuel of the future.

(c) Paralysis

Paralysis is an old technology with a new lease of life. In its simplest form it involves heating the biomass to drive off the volatile matter, leaving behind the black residue we know as charcoal. This has double the energy density of the original material. This means that charcoal, which is half the weight of the original biomass, contains the same amount of energy – making the fuel more transportable. The charcoal also burns at a much higher temperature than the original biomass, making it more useful for manufacturing processes. More sophisticated Paralysis techniques have been developed recently to collect the volatiles that are otherwise lost to the system. The collected volatiles produce a gas rich in hydrogen (a potential fuel) and carbon monoxide. These compounds, if desired, can be synthesized into methane, methanol and other hydrocarbons. 'Flash' Paralysis can be used to produce bio-crude – a combustible fuel.

(d) Digestion

Biomass digestion works by the action of anaerobic bacteria. These microorganisms usually live at the bottom of swamps or in other places where there is no air, consuming dead organic matter to produce, among other things, methane and hydrogen.

We can put these bacteria to work for us. By feeding organic matter such as animal dung or human sewage into tanks – called digesters - and adding bacteria, we can collect the emitted gas to use as an energy source. This can be a very efficient means of extracting usable energy from such biomass – up to two-thirds of the fuel energy of the animal dung is recovered.

Another, related, technique is to collect gas from landfill sites. A large proportion of household biomass waste, such as kitchen scraps, lawn clippings and pruning, ends up at the local tip. Over a period of several decades, anaerobic bacteria are at work at the bottom of such tips, steadily decomposing the organic matter and emitting methane. The gas can be extracted and used by 'capping' a landfill site with an impervious layer of clay and then inserting perforated pipes that collect the gas and bring it to the surface.

(e) Fermentation

Like many of the other processes described here, fermentation isn't a new idea. For centuries, people have used yeasts and other microorganisms to ferment the sugar of various plants into ethanol. Producing fuel from biomass by fermentation is just an extension of this old process, although a wider range of plant material can now be used, from sugar cane to wood fiber. For instance, the waste from a wheat mill in New South Wales has been used to produce ethanol through fermentation. This is then mixed with diesel to produce 'dishelm', a product used by some trucks and buses in Sydney and Canberra.

An elaborated discussion on Digestion and Gasification, which are the major ways employed in India, are explained in subsequent sections.

Anaerobic Digestion

Anaerobic digestion is a collection of processes by which microorganisms break down biodegradable material in the absence of oxygen. The process is used for industrial or domestic purposes to manage waste or to produce fuels. Much of the fermentation used industrially to produce food and drink products, as well as home fermentation, uses anaerobic digestion.

Anaerobic digestion occurs naturally in some soils and in lake and oceanic basin sediments, where it is usually referred to as "anaerobic activity". This is the source of marsh gas methane as discovered by Volta in 1776.

The digestion process begins with bacterial hydrolysis of the input materials. Insoluble organic polymers, such as carbohydrates, are broken down to soluble derivatives that become available for other bacteria. Acidogenic bacteria then convert the sugars and amino acids into carbon dioxide, hydrogen, ammonia, and organic acids. These bacteria convert these resulting organic acids into acetic acid, along with additional ammonia, hydrogen, and carbon dioxide. Finally, methanogens convert these products to methane and carbon dioxide. The methanogenic archaea populations play an indispensable role in anaerobic wastewater treatments.

Anaerobic digestion is used as part of the process to treat biodegradable waste and sewage sludge. As part of an integrated waste management system, anaerobic digestion reduces the emission of landfill gas into the atmosphere. Anaerobic digesters can also be fed with purpose-grown energy crops, such as maize.

Anaerobic digestion is widely used as a source of renewable energy. The process produces a biogas, consisting of methane, carbon dioxide and traces of other 'contaminant' gases. This biogas can be used directly as fuel, in combined heat and power gas engines or upgraded to natural gas-quality biomethane. The nutrient-rich digestate also produced can be used as fertilizer.

With the re-use of waste as a resource and new technological approaches that have lowered capital costs, anaerobic digestion has in recent years received increased attention among governments in a number of countries, among these the United Kingdom (2011), Germany and Denmark (2011).

Process

Many microorganisms affect anaerobic digestion, including acetic acid-forming bacteria (acetogens) and methane-forming archaea (methanogens). These organisms promote a number of chemical processes in converting the biomass to biogas.

Gaseous oxygen is excluded from the reactions by physical containment. Anaerobes utilize electron acceptors from sources other than oxygen gas. These acceptors can be the organic material itself or may be supplied by inorganic oxides from within the input material. When the oxygen source in an anaerobic system is derived from the organic material itself, the 'intermediate' end products are primarily alcohols, aldehydes, and organic acids, plus carbon dioxide. In the presence of specialised methanogens, the intermediates are converted to the 'final' end products of methane, carbon dioxide, and trace levels of hydrogen sulfide. In an anaerobic system, the majority of the chemical energy contained within the starting material is released by methanogenic bacteria as methane.

Populations of anaerobic microorganisms typically take a significant period of time to establish themselves to be fully effective. Therefore, common practice is to introduce anaerobic microorganisms from materials with existing populations, a process known as "seeding" the digesters, typically accomplished with the addition of sewage sludge or cattle slurry.

Process Stages

The four key stages of anaerobic digestion involve hydrolysis, acidogenesis, acetogenesis and methanogenesis. The overall process can be described by the chemical reaction, where organic material such as glucose is biochemically digested into carbon dioxide (CO_2) and methane (CH_4) by the anaerobic microorganisms.

$$C_6H_{12}O_6 \rightarrow 3CO_2 + 3CH_4$$

- Hydrolysis:- In most cases, biomass is made up of large organic polymers. For the bacteria in anaerobic digesters to access the energy potential of the material, these chains must first be broken down into their smaller constituent parts. These constituent parts, or monomers, such as sugars, are readily available to other bacteria. The process of breaking these chains and dissolving the smaller molecules into solution is called hydrolysis. Therefore, hydrolysis of these high-molecular-weight polymeric components is the necessary first step in anaerobic digestion. Through hydrolysis the complex organic molecules are broken down into simple sugars, amino acids, and fatty acids.

  Acetate and hydrogen produced in the first stages can be used directly by methanogens. Other molecules, such as volatile fatty acids (VFAs) with a chain length greater than that of acetate must first be catabolised into compounds that can be directly used by methanogens.

- Acidogenesis:- The biological process of acidogenesis results in further breakdown of the remaining components by acidogenic (fermentative) bacteria. Here, VFAs are created, along with ammonia, carbon dioxide, and hydrogen sulfide, as well as other byproducts. The process of acidogenesis is similar to the way milk sours.

- Acetogenesis:- The third stage of anaerobic digestion is acetogenesis. Here, simple molecules created through the acidogenesis phase are further digested by acetogens to produce largely acetic acid, as well as carbon dioxide and hydrogen.

- Methanogenesis:- The terminal stage of anaerobic digestion is the biological process of methanogenesis. Here, methanogens use the intermediate products of the preceding stages and convert them into methane, carbon dioxide, and water. These components make up the majority of the biogas emitted from the system. Methanogenesis is sensitive to both high and low pHs and occurs between pH 6.5 and pH 8. The remaining, indigestible material the microbes cannot use and any dead bacterial remains constitute the digestate.

Configuration

Anaerobic digesters can be designed and engineered to operate using a number of different configurations and can be categorized into batch vs. continuous process mode, mesophilic vs. thermophilic temperature conditions, high vs. low portion of solids, and single stage vs. multistage processes. More initial build money and a larger volume of the batch digester is needed to handle the same amount of waste as a continuous process digester. Higher heat energy is demanded in a thermophilic system compared to a mesophilic system and has a larger gas output capacity and higher methane gas content. For solids content, low will handle up to 15% solid content. Above this level is considered high solids content and can also be known as dry digestion. In a single stage process, one reactor houses the four anaerobic digestion steps. A multistage process utilizes two or more reactors for digestion to separate the methanogenesis and hydrolysis phases.

Batch or Continuous

Anaerobic digestion can be performed as a batch process or a continuous process. In a batch system, biomass is added to the reactor at the start of the process. The reactor is then sealed for the duration of the process. In its simplest form batch processing needs inoculation with already processed material to start the anaerobic digestion. In a typical scenario, biogas production will be formed with a normal distribution pattern over time. Operators can use this fact to determine when they believe the process of digestion of the organic matter has completed. There can be severe odour issues if a batch reactor is opened and emptied before the process is well completed. A more advanced type of batch approach has limited the odour issues by integrating anaerobic digestion with in-vessel composting. In this approach inoculation takes place through the use of recirculated degasified percolate. After anaerobic digestion has completed, the biomass is kept in the reactor which is then used for in-vessel composting before it is opened As the batch digestion is simple and requires less equipment and lower levels of design work, it is typically a cheaper form of digestion. Using more than one batch reactor at a plant can ensure constant production of biogas.

In continuous digestion processes, organic matter is constantly added (continuous complete mixed) or added in stages to the reactor (continuous plug flow; first in – first out). Here, the end products are constantly or periodically removed, resulting in constant production of biogas. A single or multiple digesters in sequence may be used. Examples of this form of anaerobic digestion include continuous stirred-tank reactors, upflow anaerobic sludge blankets, expanded granular sludge beds and internal circulation reactors.

Temperature

The two conventional operational temperature levels for anaerobic digesters determine the species of methanogens in the digesters:

- *Mesophilic* digestion takes place optimally around 30 to 38 °C, or at ambient temperatures between 20 and 45 °C, where mesophiles are the primary microorganism present.
- *Thermophilic* digestion takes place optimally around 49 to 57 °C, or at elevated temperatures up to 70 °C, where thermophiles are the primary microorganisms present.

A limit case has been reached in Bolivia, with anaerobic digestion in temperature working conditions of less than 10 °C. The anaerobic process is very slow, taking more than three times the normal mesophilic time process. In experimental work at University of Alaska Fairbanks, a 1,000 litre digester using psychrophiles harvested from "mud from a frozen lake in Alaska" has produced 200–300 litres of methane per day, about 20 to 30% of the output from digesters in warmer climates. Mesophilic species outnumber thermophiles, and they are also more tolerant to changes in environmental conditions than thermophiles. Mesophilic systems are, therefore, considered to be more stable than thermophilic digestion systems. In contrast, while thermophilic digestion systems are considered less stable, their energy input is higher, with more biogas being removed from the organic matter in an equal amount of time. The increased temperatures facilitate faster reaction rates, and thus faster gas yields. Operation at higher temperatures facilitates greater pathogen reduction of the digestate. In countries where legislation, such as the Animal By-Products Regulations in the European Union, requires digestate to meet certain levels of pathogen reduction there may be a benefit to using thermophilic temperatures instead of mesophilic.

Additional pre-treatment can be used to reduce the necessary retention time to produce biogas. For example, certain processes shred the substrates to increase the surface area or use a thermal pretreatment stage (such as pasteurisation) to significantly enhance the biogas output. The pasteurisation process can also be used to reduce the pathogenic concentration in the digesate leaving the anaerobic digester. Pasteurisation may be achieved by heat treatment combined with maceration of the solids.

Solids Content

In a typical scenario, three different operational parameters are associated with the solids content of the feedstock to the digesters:

- High solids (dry—stackable substrate).
- High solids (wet—pumpable substrate).
- Low solids (wet—pumpable substrate).

High solids (dry) digesters are designed to process materials with a solids content between 25 and 40%. Unlike wet digesters that process pumpable slurries, high solids (dry – stackable substrate) digesters are designed to process solid substrates without the addition of water. The primary styles of dry digesters are continuous vertical plug flow and batch tunnel horizontal digesters. Continuous vertical plug flow digesters are upright, cylindrical tanks where feedstock is continuously fed into the top of the digester, and flows downward by gravity during digestion. In batch tunnel digesters, the feedstock is deposited in tunnel-like chambers with a gas-tight door. Neither approach has mixing inside the digester. The amount of pretreatment, such as contaminant removal, depends both upon the nature of the waste streams being processed and the desired quality of the digestate. Size reduction (grinding) is beneficial in continuous vertical systems, as it accelerates digestion, while batch systems avoid grinding and instead require structure (e.g. yard waste) to reduce compaction of the stacked pile. Continuous vertical dry digesters have a smaller footprint due to the shorter effective retention time and vertical design. Wet digesters can be designed to operate in either a high-solids content, with a total suspended solids (TSS) concentration greater than ~20%, or a low-solids concentration less than ~15%.

High solids (wet) digesters process a thick slurry that requires more energy input to move and process the feedstock. The thickness of the material may also lead to associated problems with abrasion. High solids digesters will typically have a lower land requirement due to the lower volumes associated with the moisture. High solids digesters also require correction of conventional performance calculations (e.g. gas production, retention time, kinetics, etc.) originally based on very dilute sewage digestion concepts, since larger fractions of the feedstock mass are potentially convertible to biogas.

Low solids (wet) digesters can transport material through the system using standard pumps that require significantly lower energy input. Low solids digesters require a larger amount of land than high solids due to the increased volumes associated with the increased liquid-to-feedstock ratio of the digesters. There are benefits associated with operation in a liquid environment, as it enables more thorough circulation of materials and contact between the bacteria and their food. This enables the bacteria to more readily access the substances on which they are feeding, and increases the rate of gas production.

Complexity

Digestion systems can be configured with different levels of complexity. In a single-stage digestion system (one-stage), all of the biological reactions occur within a single, sealed reactor or holding tank. Using a single stage reduces construction costs, but results in less control of the reactions occurring within the system. Acidogenic bacteria, through the production of acids, reduce the pH of the tank. Methanogenic bacteria, as outlined earlier, operate in a strictly defined pH range. Therefore, the biological reactions of the different species in a single-stage reactor can be in direct competition with each other. Another one-stage reaction system is an anaerobic lagoon. These lagoons are pond-like, earthen basins used for the treatment and long-term storage of manures. Here the anaerobic reactions are contained within the natural anaerobic sludge contained in the pool.

In a two-stage digestion system (multistage), different digestion vessels are optimised to bring maximum control over the bacterial communities living within the digesters. Acidogenic bacteria produce organic acids and more quickly grow and reproduce than methanogenic bacteria. Methanogenic bacteria require stable pH and temperature to optimise their performance.

Under typical circumstances, hydrolysis, acetogenesis, and acidogenesis occur within the first reaction vessel. The organic material is then heated to the required operational temperature (either mesophilic or thermophilic) prior to being pumped into a methanogenic reactor. The initial hydrolysis or acidogenesis tanks prior to the methanogenic reactor can provide a buffer to the rate at which feedstock is added. Some European countries require a degree of elevated heat treatment to kill harmful bacteria in the input waste. In this instance, there may be a pasteurisation or sterilisation stage prior to digestion or between the two digestion tanks. Notably, it is not possible to completely isolate the different reaction phases, and often some biogas is produced in the hydrolysis or acidogenesis tanks.

Residence Time

The residence time in a digester varies with the amount and type of feed material, and with the configuration of the digestion system. In a typical two-stage mesophilic digestion, residence time varies between 15 and 40 days, while for a single-stage thermophilic digestion, residence times is normally faster and takes around 14 days. The plug-flow nature of some of these systems will mean the full degradation of the material may not have been realised in this timescale. In this event, digestate exiting the system will be darker in colour and will typically have more odour.

In the case of an upflow anaerobic sludge blanket digestion (UASB), hydraulic residence times can be as short as 1 hour to 1 day, and solid retention times can be up to 90 days. In this manner, a UASB system is able to separate solids and hydraulic retention times with the use of a sludge blanket. Continuous digesters have mechanical or hydraulic devices, depending on the level of solids in the material, to mix the contents, enabling the bacteria and the food to be in contact. They also allow excess material to be continuously extracted to maintain a reasonably constant volume within the digestion tanks.

Inhibition

The anaerobic digestion process can be inhibited by several compounds, affecting one or more of the bacterial groups responsible for the different organic matter degradation steps. The degree of the inhibition depends, among other factors, on the concentration of the inhibitor in the digester. Potential inhibitors are ammonia, sulfide, light metal ions (Na, K, Mg, Ca, Al), heavy metals, some organics (chlorophenols, halogenated aliphatics, N-substituted aromatics, long chain fatty acids), etc.

Left: Farm-based maize silage digester located near Neumünster in Germany, 2007 - the green, inflatable biogas holder is shown on top of the digester. *Right:* Two-stage, low solids, UASB digestion component of a mechanical biological treatment system near Tel Aviv; the process water is seen in balance tank and sequencing batch reactor, 2005.

Feedstocks

Anaerobic lagoon and generators at the Cal Poly Dairy, United States.

The most important initial issue when considering the application of anaerobic digestion systems is the feedstock to the process. Almost any organic material can be processed with anaerobic digestion; however, if biogas production is the aim, the level of putrescibility is the key factor in its successful application. The more putrescible (digestible) the material, the higher the gas yields possible from the system.

Feedstocks can include biodegradable waste materials, such as waste paper, grass clippings, leftover food, sewage, and animal waste. Woody wastes are the exception, because they are largely unaffected by digestion, as most anaerobes are unable to degrade lignin. Xylophalgeous anaerobes (lignin consumers) or using high temperature pretreatment, such as pyrolysis, can be used to break down the lignin. Anaerobic digesters can also be fed with specially grown energy crops, such as silage, for dedicated biogas production. In Germany and continental Europe, these facilities are referred to as "biogas" plants. A codigestion or cofermentation plant is typically an agricultural anaerobic digester that accepts two or more input materials for simultaneous digestion.

The length of time required for anaerobic digestion depends on the chemical complexity of the material. Material rich in easily digestible sugars breaks down quickly where as intact lignocellulosic material rich in cellulose and hemicellulose polymers can take much longer to break down. Anaerobic microorganisms are generally unable to break down lignin, the recalcitrant aromatic component of biomass.

Anaerobic digesters were originally designed for operation using sewage sludge and manures. Sewage and manure are not, however, the material with the most potential for anaerobic digestion, as the biodegradable material has already had much of the energy content taken out by the animals that produced it. Therefore, many digesters operate with codigestion of two or more types of feedstock. For example, in a farm-based digester that uses dairy manure as the primary feedstock, the gas production may be significantly increased by adding a second feedstock, e.g., grass and corn (typical on-farm feedstock), or various organic byproducts, such as slaughterhouse waste, fats, oils and grease from restaurants, organic household waste, etc. (typical off-site feedstock).

Digesters processing dedicated energy crops can achieve high levels of degradation and biogas production. Slurry-only systems are generally cheaper, but generate far less energy than those using crops, such as maize and grass silage; by using a modest amount of crop material (30%), an anaerobic digestion plant can increase energy output tenfold for only three times the capital cost, relative to a slurry-only system.

Moisture Content

A second consideration related to the feedstock is moisture content. Dryer, stackable substrates, such as food and yard waste, are suitable for digestion in tunnel-like chambers. Tunnel-style systems typically have near-zero wastewater discharge, as well, so this style of system has advantages where the discharge of digester liquids are a liability. The wetter the material, the more suitable it will be to handling with standard pumps instead of energy-intensive concrete pumps and physical means of movement. Also, the wetter the material, the more volume and area it takes up relative to the levels of gas produced. The moisture content of the target feedstock will also affect what type of system is applied to its treatment. To use a high-solids anaerobic digester for dilute feedstocks, bulking agents, such as compost, should be applied to increase the solids content of the input material. Another key consideration is the carbon:nitrogen ratio of the input material. This ratio is the balance of food a microbe requires to grow; the optimal C:N ratio is 20–30:1. Excess N can lead to ammonia inhibition of digestion.

Contamination

The level of contamination of the feedstock material is a key consideration. If the feedstock to the digesters has significant levels of physical contaminants, such as plastic, glass, or metals, then processing to remove the contaminants will be required for the material to be used. If it is not removed, then the digesters can be blocked and will not function efficiently. It is with this understanding that mechanical biological treatment plants are designed. The higher the level of pre-treatment a feedstock requires, the more processing machinery will be required, and, hence, the project will have higher capital costs.

After sorting or screening to remove any physical contaminants from the feedstock, the material is often shredded, minced, and mechanically or hydraulically pulped to increase the surface area available to microbes in the digesters and, hence, increase the speed of digestion. The maceration of solids can be achieved by using a chopper pump to transfer the feedstock material into the air-tight digester, where anaerobic treatment takes place.

Substrate Composition

Substrate composition is a major factor in determining the methane yield and methane production rates from the digestion of biomass. Techniques to determine the compositional characteristics of the feedstock are available, while parameters such as solids, elemental, and organic analyses are important for digester design and operation.

Applications

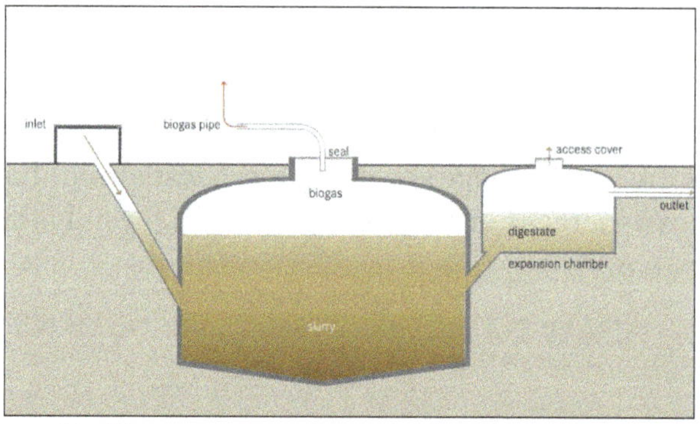

Schematic of an anaerobic digester as part of a sanitation system. It produces a digested slurry (digestate) that can be used as a fertilizer, and biogas that can be used for energy.

Using anaerobic digestion technologies can help to reduce the emission of greenhouse gases in a number of key ways:

- Replacement of fossil fuels.
- Reducing or eliminating the energy footprint of waste treatment plants.
- Reducing methane emission from landfills.
- Displacing industrially produced chemical fertilizers.
- Reducing vehicle movements.

- Reducing electrical grid transportation losses.
- Reducing usage of LP Gas for cooking.

Waste and Wastewater Treatment

Anaerobic digestion is particularly suited to organic material, and is commonly used for industrial effluent, wastewater and sewage sludge treatment. Anaerobic digestion, a simple process, can greatly reduce the amount of organic matter which might otherwise be destined to be dumped at sea, dumped in landfills, or burnt in incinerators.

Pressure from environmentally related legislation on solid waste disposal methods in developed countries has increased the application of anaerobic digestion as a process for reducing waste volumes and generating useful byproducts. It may either be used to process the source-separated fraction of municipal waste or alternatively combined with mechanical sorting systems, to process residual mixed municipal waste. These facilities are called mechanical biological treatment plants.

If the putrescible waste processed in anaerobic digesters were disposed of in a landfill, it would break down naturally and often anaerobically. In this case, the gas will eventually escape into the atmosphere. As methane is about 20 times more potent as a greenhouse gas than carbon dioxide, this has significant negative environmental effects.

In countries that collect household waste, the use of local anaerobic digestion facilities can help to reduce the amount of waste that requires transportation to centralized landfill sites or incineration facilities. This reduced burden on transportation reduces carbon emissions from the collection vehicles. If localized anaerobic digestion facilities are embedded within an electrical distribution network, they can help reduce the electrical losses associated with transporting electricity over a national grid.

Power Generation

In developing countries, simple home and farm-based anaerobic digestion systems offer the potential for low-cost energy for cooking and lighting. From 1975, China and India have both had large, government-backed schemes for adaptation of small biogas plants for use in the household for cooking and lighting. At present, projects for anaerobic digestion in the developing world can gain financial support through the United Nations Clean Development Mechanism if they are able to show they provide reduced carbon emissions.

Methane and power produced in anaerobic digestion facilities can be used to replace energy derived from fossil fuels, and hence reduce emissions of greenhouse gases, because the carbon in biodegradable material is part of a carbon cycle. The carbon released into the atmosphere from the combustion of biogas has been removed by plants for them to grow in the recent past, usually within the last decade, but more typically within the last growing season. If the plants are regrown, taking the carbon out of the atmosphere once more, the system will be carbon neutral. In contrast, carbon in fossil fuels has been sequestered in the earth for many millions of years, the combustion of which increases the overall levels of carbon dioxide in the atmosphere.

Biogas from sewage sludge treatment is sometimes used to run a gas engine to produce electrical power, some or all of which can be used to run the sewage works. Some waste heat from the engine is then used to heat the digester. The waste heat is, in general, enough to heat the digester to the required temperatures. The power potential from sewage works is limited – in the UK, there are about 80 MW total of such generation, with the potential to increase to 150 MW, which is insignificant compared to the average power demand in the UK of about 35,000 MW. The scope for biogas generation from nonsewage waste biological matter – energy crops, food waste, abattoir waste, etc. - is much higher, estimated to be capable of about 3,000 MW. Farm biogas plants using animal waste and energy crops are expected to contribute to reducing CO_2 emissions and strengthen the grid, while providing UK farmers with additional revenues.

Some countries offer incentives in the form of, for example, feed-in tariffs for feeding electricity onto the power grid to subsidize green energy production.

In Oakland, California at the East Bay Municipal Utility District's main wastewater treatment plant (EBMUD), food waste is currently codigested with primary and secondary municipal wastewater solids and other high-strength wastes. Compared to municipal wastewater solids digestion alone, food waste codigestion has many benefits. Anaerobic digestion of food waste pulp from the EBMUD food waste process provides a higher normalized energy benefit, compared to municipal wastewater solids: 730 to 1,300 kWh per dry ton of food waste applied compared to 560 to 940 kWh per dry ton of municipal wastewater solids applied.

Grid Injection

Biogas grid-injection is the injection of biogas into the natural gas grid. The raw biogas has to be previously upgraded to biomethane. This upgrading implies the removal of contaminants such as hydrogen sulphide or siloxanes, as well as the carbon dioxide. Several technologies are available for this purpose, being the most widely implemented the pressure swing adsorption (PSA), water or amine scrubbing (absorption processes) and, in the last years, membrane separation. As an alternative, the electricity and the heat can be used for on-site generation, resulting in a reduction of losses in the transportation of energy. Typical energy losses in natural gas transmission systems range from 1–2%, whereas the current energy losses on a large electrical system range from 5–8%.

In October 2010, Didcot Sewage Works became the first in the UK to produce biomethane gas supplied to the national grid, for use in up to 200 homes in Oxfordshire. By 2017, UK electricity firm Ecotricity plan to have digester fed by locally sourced grass fueling 6000 homes

Vehicle Fuel

After upgrading with the above-mentioned technologies, the biogas (transformed into biomethane) can be used as vehicle fuel in adapted vehicles. This use is very extensive in Sweden, where over 38,600 gas vehicles exist, and 60% of the vehicle gas is biomethane generated in anaerobic digestion plants.

Fertiliser and Soil Conditioner

The solid, fibrous component of the digested material can be used as a soil conditioner to increase the organic content of soils. Digester liquor can be used as a fertiliser to supply vital nutrients to soils instead of chemical fertilisers that require large amounts of energy to produce and transport. The use of manufactured fertilisers is, therefore, more carbon-intensive than the use of anaerobic digester liquor fertiliser. In countries such as Spain, where many soils are organically depleted, the markets for the digested solids can be equally as important as the biogas.

Cooking Gas

By using a bio-digester, which produces the bacteria required for decomposing, cooking gas is generated. The organic garbage like fallen leaves, kitchen waste, food waste etc. are fed into a crusher unit, where the mixture is conflated with a small amount of water. The mixture is then fed into the bio-digester, where the bacteria decomposes it to produce cooking gas. This gas is piped to kitchen stove. A 2 cubic meter bio-digester can produce 2 cubic meter of cooking gas. This is equivalent to 1 kg of LPG. The notable advantage of using a bio-digester is the sludge which is a rich organic manure.

Products

The three principal products of anaerobic digestion are biogas, digestate, and water.

Biogas

Typical composition of biogas		
Compound	**Formula**	**%**
Methane	CH_4	50–75
Carbon dioxide	CO_2	25–50
Nitrogen	N_2	0–10
Hydrogen	H_2	0–1
Hydrogen sulfide	H_2S	0–3
Oxygen	O_2	0–0

Biogas is the ultimate waste product of the bacteria feeding off the input biodegradable feedstock (the methanogenesis stage of anaerobic digestion is performed by archaea (a micro-organism on a distinctly different branch of the phylogenetic tree of life to bacteria), and is mostly methane and carbon dioxide, with a small amount hydrogen and trace hydrogen sulfide. (As-produced, biogas also contains water vapor, with the fractional water vapor volume a function of biogas temperature). Most of the biogas is produced during the middle of the digestion, after the bacterial population has grown, and tapers off as the putrescible material is exhausted. The gas is normally stored on top of the digester in an inflatable gas bubble or extracted and stored next to the facility in a gas holder.

The methane in biogas can be burned to produce both heat and electricity, usually with a reciprocating engine or microturbine often in a cogeneration arrangement where the electricity and waste heat generated are used to warm the digesters or to heat buildings. Excess electricity can be sold to suppliers or put into the local grid. Electricity produced by anaerobic digesters is considered to be renewable energy and may attract subsidies. Biogas does not contribute to increasing atmospheric carbon dioxide concentrations because the gas is not released directly into the atmosphere and the carbon dioxide comes from an organic source with a short carbon cycle.

Biogas may require treatment or 'scrubbing' to refine it for use as a fuel. Hydrogen sulfide, a toxic product formed from sulfates in the feedstock, is released as a trace component of the biogas. National environmental enforcement agencies, such as the U.S. Environmental Protection Agency or the English and Welsh Environment Agency, put strict limits on the levels of gases containing hydrogen sulfide, and, if the levels of hydrogen sulfide in the gas are high, gas scrubbing and cleaning equipment (such as amine gas treating) will be needed to process the biogas to within regionally accepted levels. Alternatively, the addition of ferrous chloride $FeCl_2$ to the digestion tanks inhibits hydrogen sulfide production.

Volatile siloxanes can also contaminate the biogas; such compounds are frequently found in household waste and wastewater. In digestion facilities accepting these materials as a component of the feedstock, low-molecular-weight siloxanes volatilise into biogas. When this gas is combusted in a gas engine, turbine, or boiler, siloxanes are converted into silicon dioxide (SiO_2), which deposits internally in the machine, increasing wear and tear. Practical and cost-effective technologies to remove siloxanes and other biogas contaminants are available at the present time. In certain applications, *in situ* treatment can be used to increase the methane purity by reducing the offgas carbon dioxide content, purging the majority of it in a secondary reactor.

In countries such as Switzerland, Germany, and Sweden, the methane in the biogas may be compressed for it to be used as a vehicle transportation fuel or input directly into the gas mains. In countries where the driver for the use of anaerobic digestion are renewable electricity subsidies, this route of treatment is less likely, as energy is required in this processing stage and reduces the overall levels available to sell.

Biogas holder with lightning protection rods and backup gas flare.

Biogas carrying pipes.

Digestate

Digestate is the solid remnants of the original input material to the digesters that the microbes cannot use. It also consists of the mineralised remains of the dead bacteria from within the digesters. Digestate can come in three forms: fibrous, liquor, or a sludge-based combination of the two fractions. In two-stage systems, different forms of digestate come from different digestion tanks. In single-stage digestion systems, the two fractions will be combined and, if desired, separated by further processing.

Acidogenic anaerobic digestate

The second byproduct (acidogenic digestate) is a stable, organic material consisting largely of lignin and cellulose, but also of a variety of mineral components in a matrix of dead bacterial cells; some plastic may be present. The material resembles domestic compost and can be used as such or to make low-grade building products, such as fibreboard. The solid digestate can also be used as feedstock for ethanol production.

The third byproduct is a liquid (methanogenic digestate) rich in nutrients, which can be used as a fertiliser, depending on the quality of the material being digested. Levels of potentially toxic elements (PTEs) should be chemically assessed. This will depend upon the quality of the original feedstock. In the case of most clean and source-separated biodegradable waste streams, the levels of PTEs will be low. In the case of wastes originating from industry, the levels of PTEs may be higher and will need to be taken into consideration when determining a suitable end use for the material.

Digestate typically contains elements, such as lignin, that cannot be broken down by the anaerobic microorganisms. Also, the digestate may contain ammonia that is phytotoxic, and may hamper the growth of plants if it is used as a soil-improving material. For these two reasons, a maturation or composting stage may be employed after digestion. Lignin and other materials are available for degradation by aerobic microorganisms, such as fungi, helping reduce the overall volume of the material for transport. During this maturation, the ammonia will be oxidized into nitrates, improving the fertility of the material and making it more suitable as a soil improver. Large composting stages are typically used by dry anaerobic digestion technologies.

Wastewater

The final output from anaerobic digestion systems is water, which originates both from the moisture content of the original waste that was treated and water produced during the microbial reactions in the digestion systems. This water may be released from the dewatering of the digestate or may be implicitly separate from the digestate.

The wastewater exiting the anaerobic digestion facility will typically have elevated levels of biochemical oxygen demand (BOD) and chemical oxygen demand (COD). These measures of the reactivity of the effluent indicate an ability to pollute. Some of this material is termed 'hard COD', meaning it cannot be accessed by the anaerobic bacteria for conversion into biogas. If this effluent were put directly into watercourses, it would negatively affect them by causing eutrophication. As such, further treatment of the wastewater is often required. This treatment will typically be an oxidation stage wherein air is passed through the water in a sequencing batch reactors or reverse osmosis unit.

History

Gas street lamp.

Reported scientific interest in the manufacturing of gas produced by the natural decomposition of organic matter dates from the 17th century, when Robert Boyle (1627-1691) and Stephen Hales (1677-1761) noted that disturbing the sediment of streams and lakes released flammable gas. In 1808 Sir Humphry Davy proved the presence of methane in the gases produced by cattle manure. In 1859 a leper colony in Bombay in India built the first anaerobic digester. In 1895, the technology was developed in Exeter, England, where a septic tank was used to generate gas for the sewer gas destructor lamp, a type of gas lighting. Also in England, in 1904, the first dual-purpose tank for both sedimentation and sludge treatment was installed in Hampton, London. In 1907, in Germany, a patent was issued for the Imhoff tank, an early form of digester.

Research on anaerobic digestion began in earnest in the 1930s.

Biofuel

A bus fueled by biodiesel.

Information on pump regarding ethanol fuel blend up to 10%, California.

A biofuel is a fuel that is produced through contemporary biological processes, such as agriculture and anaerobic digestion, rather than a fuel produced by geological processes such as those involved in the formation of fossil fuels, such as coal and petroleum, from prehistoric biological matter.

Biofuels can be derived directly from plants, or indirectly from agricultural, commercial, domestic, and/or industrial wastes. Renewable biofuels generally involve contemporary carbon fixation, such as those that occur in plants or microalgae through the process of photosynthesis. Other renewable biofuels are made through the use or conversion of biomass (referring to recently living organisms, most often referring to plants or plant-derived materials). This biomass can be converted to convenient energy-containing substances in three different ways: thermal conversion, chemical conversion, and biochemical conversion. This biomass conversion can result in fuel in solid, liquid, or gas form. This new biomass can also be used directly for biofuels.

Bioethanol is an alcohol made by fermentation, mostly from carbohydrates produced in sugar or starch crops such as corn, sugarcane, or sweet sorghum. Cellulosic biomass, derived from non-food sources, such as trees and grasses, is also being developed as a feedstock for ethanol production. Ethanol can be used as a fuel for vehicles in its pure form, but it is usually used as a gasoline additive to increase octane and improve vehicle emissions. Bioethanol is widely used in the USA and in Brazil. Current plant design does not provide for converting the lignin portion of plant raw materials to fuel components by fermentation.

Biodiesel can be used as a fuel for vehicles in its pure form, but it is usually used as a diesel additive to reduce levels of particulates, carbon monoxide, and hydrocarbons from diesel-powered vehicles. Biodiesel is produced from oils or fats using transesterification and is the most common biofuel in Europe.

In 2010, worldwide biofuel production reached 105 billion liters (28 billion gallons US), up 17% from 2009, and biofuels provided 2.7% of the world's fuels for road transport. Global ethanol fuel production reached 86 billion liters (23 billion gallons US) in 2010, with the United States and Brazil as the world's top producers, accounting together for about 90% of global production. The world's largest biodiesel producer is the European Union, accounting for 53% of all biodiesel production in 2010. As of 2011, mandates for blending biofuels exist in 31 countries at the national level and in 29 states or provinces. The International Energy Agency has a goal for biofuels to meet more than a quarter of world demand for transportation fuels by 2050 to reduce dependence on petroleum and coal. The production of biofuels also led into a flourishing automotive industry, where by 2010, 79% of all cars produced in Brazil were made with a hybrid fuel system of bioethanol and gasoline.

There are various social, economic, environmental and technical issues relating to biofuels production and use, which have been debated in the popular media and scientific journals. These include: the effect of moderating oil prices, the "food vs fuel" debate, poverty reduction potential, carbon emissions levels, sustainable biofuel production, deforestation and soil erosion, loss of biodiversity, impact on water resources, rural social exclusion and injustice, shantytown migration, rural unskilled unemployment, and nitrogen dioxide (NO_2) emissions.

Liquid Fuels for Transportation

Most transportation fuels are liquids, because vehicles usually require high energy density. This occurs naturally in liquids and solids. High energy density can also be provided by an internal combustion engine. These engines require clean-burning fuels. The fuels that are easiest to burn cleanly are typically liquids and gases. Thus, liquids meet the requirements of being both energy-dense and clean-burning. In addition, liquids (and gases) can be pumped, which means handling is easily mechanized, and thus less laborious.

First-generation Biofuels

"First-generation" or conventional biofuels are made from sugar, starch, or vegetable oil.

Ethanol

Neat ethanol on the left (A), gasoline on the right (G) at a filling station in Brazil.

Biologically produced alcohols, most commonly ethanol, and less commonly propanol and butanol, are produced by the action of microorganisms and enzymes through the fermentation of sugars or starches (easiest), or cellulose (which is more difficult). Biobutanol (also called biogasoline) is often claimed to provide a direct replacement for gasoline, because it can be used directly in a gasoline engine.

U.S. President George W. Bush looks at sugar cane, a source of biofuel, with Brazilian President Luiz Inácio Lula da Silva during a tour on biofuel technology at Petrobras in São Paulo, Brazil, 9 March 2007.

Ethanol fuel is the most common biofuel worldwide, particularly in Brazil. Alcohol fuels are produced by fermentation of sugars derived from wheat, corn, sugar beets, sugar cane, molasses and any sugar or starch from which alcoholic beverages such as whiskey, can be made (such as potato and fruit waste, etc.). The ethanol production methods used are enzyme digestion (to release sugars from stored starches), fermentation of the sugars, distillation and drying. The distillation process requires significant energy input for heat (sometimes unsustainable natural gas fossil fuel, but cellulosic biomass such as bagasse, the waste left after sugar cane is pressed to extract its juice, is the most common fuel in Brazil, while pellets, wood chips and also waste heat are more common in Europe) Waste steam fuels ethanol factory – where waste heat from the factories also is used in the district heating grid.

Ethanol can be used in petrol engines as a replacement for gasoline; it can be mixed with gasoline to any percentage. Most existing car petrol engines can run on blends of up to 15% bioethanol with petroleum/gasoline. Ethanol has a smaller energy density than that of gasoline; this means it takes more fuel (volume and mass) to produce the same amount of work. An advantage of ethanol (CH_3CH_2OH) is that it has a higher octane rating than ethanol-free gasoline available at roadside gas stations, which allows an increase of an engine's compression ratio for increased thermal efficiency. In high-altitude (thin air) locations, some states mandate a mix of gasoline and ethanol as a winter oxidizer to reduce atmospheric pollution emissions.

Ethanol is also used to fuel bioethanol fireplaces. As they do not require a chimney and are "flue-less", bioethanol fires are extremely useful for newly built homes and apartments without a flue. The downsides to these fireplaces is that their heat output is slightly less than electric heat or gas fires, and precautions must be taken to avoid carbon monoxide poisoning.

Corn-to-ethanol and other food stocks has led to the development of cellulosic ethanol. According to a joint research agenda conducted through the US Department of Energy, the fossil energy ratios (FER) for cellulosic ethanol, corn ethanol, and gasoline are 10.3, 1.36, and 0.81, respectively.

Ethanol has roughly one-third lower energy content per unit of volume compared to gasoline. This is partly counteracted by the better efficiency when using ethanol (in a long-term test of more than 2.1 million km, the BEST project found FFV vehicles to be 1-26 % more energy efficient than petrol cars, but the volumetric consumption increases by approximately 30%, so more fuel stops are required).

With current subsidies, ethanol fuel is slightly cheaper per distance traveled in the United States.

Biodiesel

Biodiesel is the most common biofuel in Europe. It is produced from oils or fats using transesterification and is a liquid similar in composition to fossil/mineral diesel. Chemically, it consists mostly of fatty acid methyl (or ethyl) esters (FAMEs). Feedstocks for biodiesel include animal fats, vegetable oils, soy, rapeseed, jatropha, mahua, mustard, flax, sunflower, palm oil, hemp, field pennycress, *Pongamia pinnata* and algae. Pure biodiesel (B100, also known as "neat" biodiesel) currently reduces emissions with up to 60% compared to diesel Second generation B100.

Biodiesel can be used in any diesel engine when mixed with mineral diesel. In some countries, manufacturers cover their diesel engines under warranty for B100 use, although Volkswagen of Germany, for example, asks drivers to check by telephone with the VW environmental services department before switching to B100. B100 may become more viscous at lower temperatures, depending on the feedstock used. In most cases, biodiesel is compatible with diesel engines from 1994 onwards, which use 'Viton' (by DuPont) synthetic rubber in their mechanical fuel injection systems. Note however, that no vehicles are certified for using pure biodiesel before 2014, as there was no emission control protocol available for biodiesel before this date.

Electronically controlled 'common rail' and 'unit injector' type systems from the late 1990s onwards may only use biodiesel blended with conventional diesel fuel. These engines have finely metered and atomized multiple-stage injection systems that are very sensitive to the viscosity of the fuel. Many current-generation diesel engines are made so that they can run on B100 without altering the engine itself, although this depends on the fuel rail design. Since biodiesel is an effective solvent and cleans residues deposited by mineral diesel, engine filters may need to be replaced more often, as the biofuel dissolves old deposits in the fuel tank and pipes. It also effectively cleans the engine combustion chamber of carbon deposits, helping to maintain efficiency. In many European countries, a 5% biodiesel blend is widely used and is available at thousands of gas stations. Biodiesel is also an oxygenated fuel, meaning it contains a reduced amount of carbon and higher hydrogen and oxygen content than fossil diesel. This improves the combustion of biodiesel and reduces the particulate emissions from unburnt carbon. However, using pure biodiesel may increase NO_x-emissions.

Biodiesel is also safe to handle and transport because it is non-toxic and biodegradable, and has a high flash point of about 300 °F (148 °C) compared to petroleum diesel fuel, which has a flash point of 125 °F (52 °C).

In the USA, more than 80% of commercial trucks and city buses run on diesel. The emerging US biodiesel market is estimated to have grown 200% from 2004 to 2005. "By the end of 2006 biodiesel production was estimated to increase fourfold [from 2004] to more than" 1 billion US gallons (3,800,000 m^3).

In France, biodiesel is incorporated at a rate of 8% in the fuel used by all French diesel vehicles. Avril Group produces under the brand Diester, a fifth of 11 million tons of biodiesel consumed annually by the European Union. It is the leading European producer of biodiesel.

Other Bioalcohols

Methanol is currently produced from natural gas, a non-renewable fossil fuel. In the future it is hoped to be produced from biomass as biomethanol. This is technically feasible, but the production is currently being postponed for concerns of Jacob S. Gibbs and Brinsley Coleberd that the economic viability is still pending. The methanol economy is an alternative to the hydrogen economy, compared to today's hydrogen production from natural gas.

Butanol (C_4H_9OH) is formed by ABE fermentation (acetone, butanol, ethanol) and experimental modifications of the process show potentially high net energy gains with butanol as the only liquid product. Butanol will produce more energy and allegedly can be burned "straight" in existing gasoline engines (without modification to the engine or car), and is less corrosive and less water-soluble than ethanol, and could be distributed via existing infrastructures. DuPont and BP are working together to help develop butanol. *E. coli* strains have also been successfully engineered to produce butanol by modifying their amino acid metabolism.

Green Diesel

Green diesel is produced through hydrocracking biological oil feedstocks, such as vegetable oils and animal fats. Hydrocracking is a refinery method that uses elevated temperatures and pressure in the presence of a catalyst to break down larger molecules, such as those found in vegetable oils, into shorter hydrocarbon chains used in diesel engines. It may also be called renewable diesel, hydrotreated vegetable oil or hydrogen-derived renewable diesel. Green diesel has the same chemical properties as petroleum-based diesel. It does not require new engines, pipelines or infrastructure to distribute and use, but has not been produced at a cost that is competitive with petroleum. Gasoline versions are also being developed. Green diesel is being developed in Louisiana and Singapore by ConocoPhillips, Neste Oil, Valero, Dynamic Fuels, and Honeywell UOP as well as Preem in Gothenburg, Sweden, creating what is known as Evolution Diesel.

Biofuel Gasoline

In 2013 UK researchers developed a genetically modified strain of Escherichia coli (E.Coli), which could transform glucose into biofuel gasoline that does not need to be blended. Later in 2013 UCLA researchers engineered a new metabolic pathway to bypass glycolysis and increase the rate of conversion of sugars into biofuel, while KAIST researchers developed a strain capable of producing short-chain alkanes, free fatty acids, fatty esters and fatty alcohols through the fatty acyl (acyl carrier protein (ACP)) to fatty acid to fatty acyl-CoA pathway *in vivo*. It is believed that in the future it will be possible to "tweak" the genes to make gasoline from straw or animal manure.

Vegetable Oil

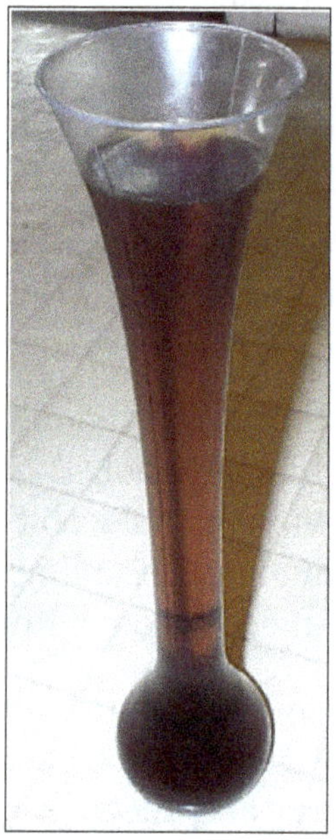

Filtered waste vegetable oil.

Walmart's truck fleet logs millions of miles each year, and the company planned to double the fleet's efficiency between 2005 and 2015. This truck is one of 15 based at Walmart's Buckeye, Arizona distribution center that was converted to run on a biofuel made from reclaimed cooking grease produced during food preparation at Walmart stores.

Straight unmodified edible vegetable oil is generally not used as fuel, but lower-quality oil has been used for this purpose. Used vegetable oil is increasingly being processed into biodiesel, or (more rarely) cleaned of water and particulates and then used as a fuel.

As with 100% biodiesel (B100), to ensure the fuel injectors atomize the vegetable oil in the correct pattern for efficient combustion, vegetable oil fuel must be heated to reduce its viscosity to that of diesel, either by electric coils or heat exchangers. This is easier in warm or temperate climates. MAN B&W Diesel, Wärtsilä, and Deutz AG, as well as a number of smaller companies, such as Elsbett, offer engines that are compatible with straight vegetable oil, without the need for after-market modifications.

Vegetable oil can also be used in many older diesel engines that do not use common rail or unit injection electronic diesel injection systems. Due to the design of the combustion chambers in indirect injection engines, these are the best engines for use with vegetable oil. This system allows the relatively larger oil molecules more time to burn. Some older engines, especially Mercedes, are driven experimentally by enthusiasts without any conversion, a handful of drivers have experienced limited success with earlier pre-"Pumpe Duse" VW TDI engines and other similar engines with direct injection. Several companies, such as Elsbett or Wolf, have developed professional conversion kits and successfully installed hundreds of them over the last decades.

Oils and fats can be hydrogenated to give a diesel substitute. The resulting product is a straight-chain hydrocarbon with a high cetane number, low in aromatics and sulfur and does not contain oxygen. Hydrogenated oils can be blended with diesel in all proportions. They have several advantages over biodiesel, including good performance at low temperatures, no storage stability problems and no susceptibility to microbial attack.

Bioethers

Bioethers (also referred to as fuel ethers or oxygenated fuels) are cost-effective compounds that act as octane rating enhancers."Bioethers are produced by the reaction of reactive iso-olefins, such as iso-butylene, with bioethanol." Bioethers are created by wheat or sugar beet. They also enhance engine performance, whilst significantly reducing engine wear and toxic exhaust emissions. Though bioethers are likely to replace petroethers in the UK, it is highly unlikely they will become a fuel in and of itself due to the low energy density. Greatly reducing the amount of ground-level ozone emissions, they contribute to air quality.

When it comes to transportation fuel there are six ether additives: dimethyl ether (DME), diethyl ether (DEE), methyl teritiary-butyl ether (MTBE), ethyl *ter*-butyl ether (ETBE), *ter*-amyl methyl ether (TAME), and *ter*-amyl ethyl ether (TAEE).

The European Fuel Oxygenates Association (EFOA) credits methyl Ttertiary-butyl ether (MTBE) and ethyl ter-butyl ether (ETBE) as the most commonly used ethers in fuel to replace lead. Ethers were introduced in Europe in the 1970s to replace the highly toxic compound. Although Europeans still use bio-ether additives, the US no longer has an oxygenate requirement therefore bio-ethers are no longer used as the main fuel additive.

Biogas

Biogas is methane produced by the process of anaerobic digestion of organic material by anaerobes. It can be produced either from biodegradable waste materials or by the use of energy crops fed into anaerobic digesters to supplement gas yields. The solid byproduct, digestate, can be used as a biofuel or a fertilizer.

Biogas can be recovered from mechanical biological treatment waste processing systems. Landfill gas, a less clean form of biogas, is produced in landfills through naturally occurring anaerobic digestion. If it escapes into the atmosphere, it is a potential greenhouse gas.

Farmers can produce biogas from manure from their cattle by using anaerobic digesters.

Syngas

Syngas, a mixture of carbon monoxide, hydrogen and other hydrocarbons, is produced by partial combustion of biomass, that is, combustion with an amount of oxygen that is not sufficient to convert the biomass completely to carbon dioxide and water. Before partial combustion, the biomass is dried, and sometimes pyrolysed. The resulting gas mixture, syngas, is more efficient than direct combustion of the original biofuel; more of the energy contained in the fuel is extracted.

Syngas may be burned directly in internal combustion engines, turbines or high-temperature fuel cells. The wood gas generator, a wood-fueled gasification reactor, can be connected to an internal combustion engine.

Syngas can be used to produce methanol, DME and hydrogen, or converted via the Fischer-Tropsch process to produce a diesel substitute, or a mixture of alcohols that can be blended into gasoline. Gasification normally relies on temperatures greater than 700 °C.

Lower-temperature gasification is desirable when co-producing biochar, but results in syngas polluted with tar.

Solid Biomass Fuels

Examples include wood, sawdust, grass trimmings, domestic refuse, charcoal, agricultural waste, nonfood energy crops, and dried manure.

When solid biomass is already in a suitable form (such as firewood), it can burn directly in a stove or furnace to provide heat or raise steam. When solid biomass is in an inconvenient form (such as sawdust, wood chips, grass, urban waste wood, agricultural residues), the typical process is to densify the biomass. This process includes grinding the raw biomass to an appropriate particulate size (known as hogfuel), which, depending on the densification type, can be from 1 to 3 cm (0.4 to 1.2 in), which is then concentrated into a fuel product. The current processes produce wood pellets, cubes, or pucks. The pellet process is most common in Europe, and is typically a pure wood product. The other types of densification are larger in size compared to a pellet and are compatible with a broad range of input feedstocks. The resulting densified fuel is easier to transport and feed into thermal generation systems, such as boilers.

Sawdust, bark and chips are already used for decades for fuel in industrial processes; examples include the pulp and paper industry and the sugar cane industry. Boilers in the range of 500,000 lb/hr of steam, and larger, are in routine operation, using grate, spreader stoker, suspension burning and fluid bed combustion. Utilities generate power, typically in the range of 5 to 50 MW, using locally available fuel. Other industries have also installed wood waste fueled boilers and dryers in areas with low-cost fuel.

One of the advantages of solid biomass fuel is that it is often a byproduct, residue or waste-product of other processes, such as farming, animal husbandry and forestry. In theory, this means fuel and food production do not compete for resources, although this is not always the case.

A problem with the combustion of solid biomass fuels is that it emits considerable amounts of pollutants, such as particulates and polycyclic aromatic hydrocarbons. Even modern pellet boilers generate much more pollutants than oil or natural gas boilers. Pellets made from agricultural residues are usually worse than wood pellets, producing much larger emissions of dioxins and chlorophenols.

A derived fuel is biochar, which is produced by biomass pyrolysis. Biochar made from agricultural waste can substitute for wood charcoal. As wood stock becomes scarce, this alternative is gaining ground. In eastern Democratic Republic of Congo, for example, biomass briquettes are being marketed as an alternative to charcoal to protect Virunga National Park from deforestation associated with charcoal production.

Second-generation (Advanced) Biofuels

Second generation biofuels, also known as advanced biofuels, are fuels that can be manufactured from various types of biomass. Biomass is a wide-ranging term meaning any source of organic carbon that is renewed rapidly as part of the carbon cycle. Biomass is derived from plant materials, but can also include animal materials.

First generation biofuels are made from the sugars and vegetable oils found in arable crops, which can be easily extracted using conventional technology. In comparison, second generation biofuels are made from lignocellulosic biomass or woody crops, agricultural residues or waste. This makes it more difficult to extract the required fuel. A series of physical and chemical treatments might be required to convert lignocellulosic biomass to liquid fuels suitable for transportation.

Sustainable Biofuels

Biofuels in the form of liquid fuels derived from plant materials are entering the market, driven mainly by the perception that they reduce climate gas emissions, and also by factors such as oil price spikes and the need for increased energy security. However, many of the biofuels that are currently being supplied have been criticised for their adverse impacts on the natural environment, food security, and land use. In 2008, the Nobel-prize winning chemist Paul J. Crutzen published findings that the release of nitrous oxide (N_2O) emissions in the production of biofuels means that overall they contribute more to global warming than the fossil fuels they replace.

The challenge is to support biofuel development, including the development of new cellulosic technologies, with responsible policies and economic instruments to help ensure that biofuel commercialization is sustainable. Responsible commercialization of biofuels represents an opportunity to enhance sustainable economic prospects in Africa, Latin America and Asia.

According to the Rocky Mountain Institute, sound biofuel production practices would not hamper food and fibre production, nor cause water or environmental problems, and would enhance soil fertility. The selection of land on which to grow the feedstocks is a critical component of the ability of biofuels to deliver sustainable solutions. A key consideration is the minimisation of biofuel competition for prime cropland.

Biofuels by Region

Bio Diesel Powered Fast Attack Craft Of Indian Navy patrolling during IFR 2016.The green bands on the vessels are indicative of the fact that the vessels are powered by bio-diesel.

There are international organizations such as IEA Bioenergy, established in 1978 by the OECD International Energy Agency (IEA), with the aim of improving cooperation and information exchange between countries that have national programs in bioenergy research, development and deployment. The UN International Biofuels Forum is formed by Brazil, China, India, Pakistan, South Africa, the United States and the European Commission. The world leaders in biofuel development and use are Brazil, the United States, France, Sweden and Germany. Russia also has 22% of world's forest, and is a big biomass (solid biofuels) supplier. In 2010, Russian pulp and paper maker, Vyborgskaya Cellulose, said they would be producing pellets that can be used in heat and electricity generation from its plant in Vyborg by the end of the year. The plant will eventually produce about 900,000 tons of pellets per year, making it the largest in the world once operational.

Biofuels currently make up 3.1% of the total road transport fuel in the UK or 1,440 million litres. By 2020, 10% of the energy used in UK road and rail transport must come from renewable sources – this is the equivalent of replacing 4.3 million tonnes of fossil oil each year. Conventional biofuels are likely to produce between 3.7 and 6.6% of the energy needed in road and rail transport, while advanced biofuels could meet up to 4.3% of the UK's renewable transport fuel target by 2020.

Air Pollution

Biofuels are different from fossil fuels in regard to greenhouse gases but are similar to fossil fuels in that biofuels contribute to air pollution. Burning produces airborne carbon particulates, carbon monoxide and nitrous oxides. The WHO estimates 3.7 million premature deaths worldwide in 2012 due to air pollution. Brazil burns significant amounts of ethanol biofuel. Gas chromatograph studies were performed of ambient air in São Paulo, Brazil, and compared to Osaka, Japan, which does not burn ethanol fuel. Atmospheric Formaldehyde was 160% higher in Brazil, and Acetaldehyde was 260% higher.

Debates Regarding the Production and use of Biofuel

There are various social, economic, environmental and technical issues with biofuel production and use, which have been discussed in the popular media and scientific journals. These include the effect of moderating oil prices, the "food vs fuel" debate, food prices, poverty reduction potential, energy ratio, energy requirements, carbon emissions levels, sustainable biofuel production, deforestation and soil erosion, loss of biodiversity, impact on water resources, the possible modifications necessary to run the engine on biofuel, as well as energy balance and efficiency. The International Resource Panel, which provides independent scientific assessments and expert advice on a variety of resource-related themes, assessed the issues relating to biofuel use in its first report *Towards sustainable production and use of resources: Assessing Biofuels*. "Assessing Biofuels" outlined the wider and interrelated factors that need to be considered when deciding on the relative merits of pursuing one biofuel over another. It concluded that not all biofuels perform equally in terms of their impact on climate, energy security and ecosystems, and suggested that environmental and social impacts need to be assessed throughout the entire life-cycle.

Another issue with biofuel use and production is the US has changed mandates many times because the production has been taking longer than expected. The Renewable Fuel Standard (RFS) set by congress for 2010 was pushed back to at best 2012 to produce 100 million gallons of pure ethanol (not blended with a fossil fuel).

Current Research

Research is ongoing into finding more suitable biofuel crops and improving the oil yields of these crops. Using the current yields, vast amounts of land and fresh water would be needed to produce enough oil to completely replace fossil fuel usage. It would require twice the land area of the US to be devoted to soybean production, or two-thirds to be devoted to rapeseed production, to meet current US heating and transportation needs.

Specially bred mustard varieties can produce reasonably high oil yields and are very useful in crop rotation with cereals, and have the added benefit that the meal left over after the oil has been pressed out can act as an effective and biodegradable pesticide.

The NFESC, with Santa Barbara-based Biodiesel Industries, is working to develop biofuels technologies for the US navy and military, one of the largest diesel fuel users in the world. A group of Spanish developers working for a company called Ecofasa announced a new biofuel made from trash. The fuel is created from general urban waste which is treated by bacteria to produce fatty acids, which can be used to make biofuels.

Ethanol Biofuels (Bioethanol)

As the primary source of biofuels in North America, many organizations are conducting research in the area of ethanol production. The National Corn-to-Ethanol Research Center (NCERC) is a research division of Southern Illinois University Edwardsville dedicated solely to ethanol-based biofuel research projects. On the federal level, the USDA conducts a large amount of research regarding ethanol production in the United States. Much of this research is targeted toward the effect of ethanol production on domestic food markets. A division of the U.S. Department of Energy, the National Renewable Energy Laboratory (NREL), has also conducted various ethanol research projects, mainly in the area of cellulosic ethanol.

Cellulosic ethanol commercialization is the process of building an industry out of methods of turning cellulose-containing organic matter into fuel. Companies, such as Iogen, POET, and Abengoa, are building refineries that can process biomass and turn it into bioethanol. Companies, such as Diversa, Novozymes, and Dyadic, are producing enzymes that could enable a cellulosic ethanol future. The shift from food crop feedstocks to waste residues and native grasses offers significant opportunities for a range of players, from farmers to biotechnology firms, and from project developers to investors.

As of 2013, the first commercial-scale plants to produce cellulosic biofuels have begun operating. Multiple pathways for the conversion of different biofuel feedstocks are being used. In the next few years, the cost data of these technologies operating at commercial scale, and their relative performance, will become available. Lessons learnt will lower the costs of the industrial processes involved.

In parts of Asia and Africa where drylands prevail, sweet sorghum is being investigated as a potential source of food, feed and fuel combined. The crop is particularly suitable for growing in arid conditions, as it only extracts one seventh of the water used by sugarcane. In India, and other places, sweet sorghum stalks are used to produce biofuel by squeezing the juice and then fermenting into ethanol.

A study by researchers at the International Crops Research Institute for the Semi-Arid Tropics (ICRISAT) found that growing sweet sorghum instead of grain sorghum could increase farmers incomes by US$40 per hectare per crop because it can provide fuel in addition to food and animal feed. With grain sorghum currently grown on over 11 million hectares (ha) in Asia and on 23.4 million ha in Africa, a switch to sweet sorghum could have a considerable economic impact.

Algae Biofuels

From 1978 to 1996, the US NREL experimented with using algae as a biofuels source in the "Aquatic Species Program". A self-published article by Michael Briggs, at the UNH Biofuels Group, offers estimates for the realistic replacement of all vehicular fuel with biofuels by using algae that have a natural oil content greater than 50%, which Briggs suggests can be grown on algae ponds at wastewater treatment plants. This oil-rich algae can then be extracted from the system and processed into biofuels, with the dried remainder further reprocessed to create ethanol. The production of algae to harvest oil for biofuels has not yet been undertaken on a commercial scale, but feasibility studies have been conducted to arrive at the above yield estimate. In addition to its projected high yield, algaculture — unlike crop-based biofuels – does not entail a decrease in food production, since it requires neither farmland nor fresh water. Many companies are pursuing algae bioreactors for various purposes, including scaling up biofuels production to commercial levels. Prof. Rodrigo E. Teixeira from the University of Alabama in Huntsville demonstrated the extraction of biofuels lipids from wet algae using a simple and economical reaction in ionic liquids.

Jatropha

Several groups in various sectors are conducting research on *Jatropha curcas*, a poisonous shrub-like tree that produces seeds considered by many to be a viable source of biofuels feedstock oil. Much of this research focuses on improving the overall per acre oil yield of Jatropha through advancements in genetics, soil science, and horticultural practices.

SG Biofuels, a San Diego-based jatropha developer, has used molecular breeding and biotechnology to produce elite hybrid seeds that show significant yield improvements over first-generation varieties. SG Biofuels also claims additional benefits have arisen from such strains, including improved flowering synchronicity, higher resistance to pests and diseases, and increased cold-weather tolerance.

Plant Research International, a department of the Wageningen University and Research Centre in the Netherlands, maintains an ongoing Jatropha Evaluation Project that examines the feasibility of large-scale jatropha cultivation through field and laboratory experiments. The Center for Sustainable Energy Farming (CfSEF) is a Los Angeles-based nonprofit research organization dedicated to jatropha research in the areas of plant science, agronomy, and horticulture. Successful exploration of these disciplines is projected to increase jatropha farm production yields by 200-300% in the next 10 years.

Fungi

A group at the Russian Academy of Sciences in Moscow, in a 2008 paper, stated they had isolated large amounts of lipids from single-celled fungi and turned it into biofuels in an economically efficient manner. More research on this fungal species, *Cunninghamella japonica*, and others, is likely to appear in the near future. The recent discovery of a variant of the fungus *Gliocladium roseum* (later renamed *Ascocoryne sarcoides*) points toward the production of so-called myco-diesel from cellulose. This organism was recently discovered in the rainforests of northern Patagonia, and has the unique capability of converting cellulose into medium-length hydrocarbons typically found in diesel fuel. Many other fungi that can degrade cellulose and other polymers have been observed to produce molecules that are currently being engineered using organisms from other kingdoms, suggesting that fungi may play a large role in the bio-production of fuels in the future (reviewed in).

Animal Gut Bacteria

Microbial gastrointestinal flora in a variety of animals have shown potential for the production of biofuels. Recent research has shown that TU-103, a strain of *Clostridium* bacteria found in Zebra feces, can convert nearly any form of cellulose into butanol fuel. Microbes in panda waste are being investigated for their use in creating biofuels from bamboo and other plant materials. There has also been substantial research into the technology of using the gut microbiomes of wood-feeding insects for the conversion of lignocellulotic material into biofuel.

Greenhouse Gas Emissions

Some scientists have expressed concerns about land-use change in response to greater demand for crops to use for biofuel and the subsequent carbon emissions. The payback period, that is, the time it will take biofuels to pay back the carbon debt they acquire due to land-use change, has been estimated to be between 100 and 1000 years, depending on the specific instance and location of land-use change. However, no-till practices combined with cover-crop practices can reduce the payback period to three years for grassland conversion and 14 years for forest conversion.

A study conducted in the Tocantis State, in northern Brazil, found that many families were cutting down forests in order to produce two conglomerates of oilseed plants, the J. curcas (JC group) and the R. communis (RC group). This region is composed of 15% Amazonian rainforest with high biodiversity, and 80% Cerrado forest with lower biodiversity. During the study, the farmers that planted the JC group released over 2193 Mg CO_2, while losing 53-105 Mg CO_2 sequestration from deforestation; and the RC group farmers released 562 Mg CO_2, while losing 48-90 Mg CO_2 to be sequestered from forest depletion. The production of these types of biofuels not only led into an increased emission of carbon dioxide, but also to lower efficiency of forests to absorb the gases that these farms were emitting. This has to do with the amount of fossil fuel the production of fuel crops involves. In addition, the intensive use of monocropping agriculture requires large amounts of water irrigation, as well as of fertilizers, herbicides and pesticides. This does not only lead to poisonous chemicals to disperse on water runoff, but also to the emission of nitrous oxide (NO_2) as a fertilizer byproduct, which is three hundred times more efficient in producing a greenhouse effect than carbon dioxide (CO_2).

Converting rainforests, peatlands, savannas, or grasslands to produce food crop–based biofuels in Brazil, Southeast Asia, and the United States creates a "biofuel carbon debt" by releasing 17 to 420 times more CO_2 than the annual greenhouse gas (GHG) reductions that these biofuels would provide by displacing fossil fuels. Biofuels made from waste biomass or from biomass grown on abandoned agricultural lands incur little to no carbon debt.

Water Use

In addition to crop growth requiring water, biofuel facilities require significant process water.

Biogas

Pipes carrying biogas (foreground), natural gas and condensate.

Biogas typically refers to a mixture of different gases produced by the breakdown of organic matter in the absence of oxygen. Biogas can be produced from raw materials such as agricultural waste, manure, municipal waste, plant material, sewage, green waste or food waste. Biogas is a renewable energy source.

Biogas can be produced by anaerobic digestion with anaerobic organisms, which digest material inside a closed system, or fermentation of biodegradable materials.

Biogas is primarily methane (CH_4) and carbon dioxide (CO_2) and may have small amounts of hydrogen sulfide (H_2S), moisture and siloxanes. The gases methane, hydrogen, and carbon monoxide (CO) can be combusted or oxidized with oxygen. This energy release allows biogas to be used as a fuel; it can be used for any heating purpose, such as cooking. It can also be used in a gas engine to convert the energy in the gas into electricity and heat.

Biogas can be compressed, the same way natural gas is compressed to CNG, and used to power motor vehicles. In the UK, for example, biogas is estimated to have the potential to replace around 17% of vehicle fuel. It qualifies for renewable energy subsidies in some parts of the world. Biogas can be cleaned and upgraded to natural gas standards, when it becomes bio-methane. Biogas is considered to be a renewable resource because its production-and-use cycle is continuous, and it generates no net carbon dioxide. Organic material grows, is converted and used and then regrows in a continually repeating cycle. From a carbon perspective, as much carbon dioxide is absorbed from the atmosphere in the growth of the primary bio-resource as is released when the material is ultimately converted to energy.

Production

Biogas production in rural Germany.

Biogas is produced as landfill gas (LFG), which is produced by the breakdown of Biodegradable waste inside a landfill due to chemical reactions and microbes, or as digested gas, produced inside an anaerobic digester. A *biogas plant* is the name often given to an anaerobic digester that treats farm wastes or energy crops. It can be produced using anaerobic digesters (air-tight tanks with different configurations). These plants can be fed with energy crops such as maize silage or biodegradable wastes including sewage sludge and food waste. During the process, the microorganisms transform biomass waste into biogas (mainly methane and carbon dioxide) and digestate. The biogas is a renewable energy that can be used for heating, electricity, and many other operations that use a reciprocating internal combustion engine, such as GE Jenbacher or Caterpillar gas engines. Other internal combustion engines such as gas turbines are suitable for the conversion of biogas into both electricity and heat. The digestate is the remaining inorganic matter that was not transformed into biogas. It can be used as an agricultural fertiliser.

There are two key processes: mesophilic and thermophilic digestion which is dependent on temperature. In experimental work at University of Alaska Fairbanks, a 1000-litre digester using psychrophiles harvested from "mud from a frozen lake in Alaska" has produced 200–300 liters of methane per day, about 20%–30% of the output from digesters in warmer climates.

Dangers

The dangers of biogas are mostly similar to those of natural gas, but with an additional risk from the toxicity of its hydrogen sulfide fraction. Biogas can be explosive when mixed in the ratio of one part biogas to 8-20 parts air. Special safety precautions have to be taken for entering an empty biogas digester for maintenance work.

It is important that a biogas system never has negative pressure as this could cause an explosion. Negative gas pressure can occur if too much gas is removed or leaked; Because of this biogas should not be used at pressures below one column inch of water, measured by a pressure gauge.

Frequent smell checks must be performed on a biogas system. If biogas is smelled anywhere windows and doors should be opened immediately. If there is a fire the gas should be shut off at the gate valve of the biogas system.

Landfill Gas

Landfill gas is produced by wet organic waste decomposing under anaerobic conditions in a biogas.

The waste is covered and mechanically compressed by the weight of the material that is deposited above. This material prevents oxygen exposure thus allowing anaerobic microbes to thrive. Biogas builds up and is slowly released into the atmosphere if the site has not been engineered to capture the gas. Landfill gas released in an uncontrolled way can be hazardous since it can become explosive when it escapes from the landfill and mixes with oxygen. The lower explosive limit is 5% methane and the upper is 15% methane.

The methane in biogas is 20 times more potent a greenhouse gas than carbon dioxide. Therefore, uncontained landfill gas, which escapes into the atmosphere may significantly contribute to the effects of global warming. In addition, volatile organic compounds (VOCs) in landfill gas contribute to the formation of photochemical smog.

Technical

Biochemical oxygen demand (BOD) is a measure of the amount of oxygen required by aerobic micro-organisms to decompose the organic matter in a sample of water. Knowing the energy density of the material being used in the biodigester as well as the BOD for the liquid discharge allows for the calculation of the daily energy output from a biodigester.

Another term related to biodigesters is effluent dirtiness, which tells how much organic material there is per unit of biogas source. Typical units for this measure are in mg BOD/litre. As an example, effluent dirtiness can range between 800–1200 mg BOD/litre in Panama.

From 1 kg of decommissioned kitchen bio-waste, 0.45 m^3 of biogas can be obtained. The price for collecting biological waste from households is approximately €70 per ton.

Composition

Typical composition of biogas		
Compound	**Formula**	**%**
Methane	CH_4	50–75
Carbon dioxide	CO_2	25–50
Nitrogen	N_2	0–10
Hydrogen	H_2	0–1
Hydrogen sulfide	H_2S	0–3
Oxygen	O_2	0–0.5

The composition of biogas varies depending upon the origin of the anaerobic digestion process. Landfill gas typically has methane concentrations around 50%. Advanced waste treatment technologies can produce biogas with 55%–75% methane, which for reactors with free liquids can be increased to 80%-90% methane using in-situ gas purification techniques. As produced, biogas contains water vapor. The fractional volume of water vapor is a function of biogas temperature; correction of measured gas volume for water vapor content and thermal expansion is easily done via simple mathematics which yields the standardized volume of dry biogas.

In some cases, biogas contains siloxanes. They are formed from the anaerobic decomposition of materials commonly found in soaps and detergents. During combustion of biogas containing siloxanes, silicon is released and can combine with free oxygen or other elements in the combustion gas. Deposits are formed containing mostly silica (SiO_2) or silicates (Si_xO_y) and can contain calcium, sulfur, zinc, phosphorus. Such *white mineral* deposits accumulate to a surface thickness of several millimeters and must be removed by chemical or mechanical means.

Practical and cost-effective technologies to remove siloxanes and other biogas contaminants are available.

For 1000 kg (wet weight) of input to a typical biodigester, total solids may be 30% of the wet weight while volatile suspended solids may be 90% of the total solids. Protein would be 20% of the volatile solids, carbohydrates would be 70% of the volatile solids, and finally fats would be 10% of the volatile solids.

Benefits of Manure Derived Biogas

High levels of methane are produced when manure is stored under anaerobic conditions. During storage and when manure has been applied to the land, nitrous oxide is also produced as a byproduct of the denitrification process. Nitrous oxide (N_2O) is 320 times more aggressive as a greenhouse gas than carbon dioxide and methane 25 times more than carbon dioxide.

By converting cow manure into methane biogas via anaerobic digestion, the millions of cattle in the United States would be able to produce 100 billion kilowatt hours of electricity, enough to power millions of homes across the United States. In fact, one cow can produce enough manure in one day to generate 3 kilowatt hours of electricity; only 2.4 kilowatt hours of electricity are needed to power a single 100-watt light bulb for one day. Furthermore, by converting cattle manure into methane biogas instead of letting it decompose, global warming gases could be reduced by 99 million metric tons or 4%.

Applications

A biogas bus in Linköping, Sweden.

Biogas can be used for electricity production on sewage works, in a CHP gas engine, where the waste heat from the engine is conveniently used for heating the digester; cooking; space heating; water heating; and process heating. If compressed, it can replace compressed natural gas for use in vehicles, where it can fuel an internal combustion engine or fuel cells and is a much more effective displacer of carbon dioxide than the normal use in on-site CHP plants.

Biogas Upgrading

Raw biogas produced from digestion is roughly 60% methane and 29% CO_2 with trace elements of H_2S; it is not of high enough quality to be used as fuel gas for machinery. The corrosive nature of H_2S alone is enough to destroy the internals of a plant.

Methane in biogas can be concentrated via a biogas upgrader to the same standards as fossil natural gas, which itself has to go through a cleaning process, and becomes *biomethane*. If the local gas network allows, the producer of the biogas may use their distribution networks. Gas must be very clean to reach pipeline quality and must be of the correct composition for the distribution network to accept. Carbon dioxide, water, hydrogen sulfide, and particulates must be removed if present.

There are four main methods of upgrading: water washing, pressure swing adsorption, selexol adsorption, and amine gas treating. In addition to these, the use of membrane separation technology for biogas upgrading is increasing, and there are already several plants operating in Europe and USA.

The most prevalent method is water washing where high pressure gas flows into a column where the carbon dioxide and other trace elements are scrubbed by cascading water running counter-flow to the gas. This arrangement could deliver 98% methane with manufacturers guaranteeing maximum 2% methane loss in the system. It takes roughly between 3% and 6% of the total energy output in gas to run a biogas upgrading system.

Biogas Gas-grid Injection

Gas-grid injection is the injection of biogas into the methane grid (natural gas grid). Injections includes biogas until the breakthrough of micro combined heat and power two-thirds of all the energy produced by biogas power plants was lost (the heat), using the grid to transport the gas to customers, the electricity and the heat can be used for on-site generation resulting in a reduction of losses in the transportation of energy. Typical energy losses in natural gas transmission systems range from 1% to 2%. The current energy losses on a large electrical system range from 5% to 8%.

Biogas in Transport

"Biogaståget Amanda" ("Amanda the Biogas Train") train near Linköping station, Sweden.

If concentrated and compressed, it can be used in vehicle transportation. Compressed biogas is becoming widely used in Sweden, Switzerland, and Germany. A biogas-powered train, named Biogaståget Amanda (The Biogas Train Amanda), has been in service in Sweden since 2005. Biogas powers automobiles. In 1974, a British documentary film titled *Sweet as a Nut* detailed the biogas production process from pig manure and showed how it fueled a custom-adapted combustion engine. In 2007, an estimated 12,000 vehicles were being fueled with upgraded biogas worldwide, mostly in Europe.

Measuring in Biogas Environments

Biogas is part of the wet gas and condensing gas (or air) category that includes mist or fog in the gas stream. The mist or fog is predominately water vapor that condenses on the sides of pipes or stacks throughout the gas flow. Biogas environments include wastewater digesters, landfills, and animal feeding operations (covered livestock lagoons).

Ultrasonic flow meters are one of the few devices capable of measuring in a biogas atmosphere. Most of thermal flow meters are unable to provide reliable data because the moisture causes steady high flow readings and continuous flow spiking, although there are single-point insertion thermal mass flow meters capable of accurately monitoring biogas flows with minimal pressure drop. They can handle moisture variations that occur in the flow stream because of daily and seasonal temperature fluctuations, and account for the moisture in the flow stream to produce a dry gas value.

Legislation

European Union

The European Union has legislation regarding waste management and landfill sites called the Landfill Directive.

Countries such as the United Kingdom and Germany now have legislation in force that provides farmers with long-term revenue and energy security.

United States

The United States legislates against landfill gas as it contains VOCs. The United States Clean Air Act and Title 40 of the Code of Federal Regulations (CFR) requires landfill owners to estimate the quantity of non-methane organic compounds (NMOCs) emitted. If the estimated NMOC emissions exceeds 50 tonnes per year, the landfill owner is required to collect the gas and treat it to remove the entrained NMOCs. Treatment of the landfill gas is usually by combustion. Because of the remoteness of landfill sites, it is sometimes not economically feasible to produce electricity from the gas.

Global Developments

United States

With the many benefits of biogas, it is starting to become a popular source of energy and is starting to be used in the United States more. In 2003, the United States consumed 147 trillion BTU of energy from "landfill gas", about 0.6% of the total U.S. natural gas consumption. Methane biogas derived from cow manure is being tested in the U.S. According to a 2008 study, collected by the *Science and Children* magazine, methane biogas from cow manure would be sufficient to produce 100 billion kilowatt hours enough to power millions of homes across America. Furthermore, methane biogas has been tested to prove that it can reduce 99 million metric tons of greenhouse gas emissions or about 4% of the greenhouse gases produced by the United States.

In Vermont, for example, biogas generated on dairy farms was included in the CVPS Cow Power program. The program was originally offered by Central Vermont Public Service Corporation as a voluntary tariff and now with a recent merger with Green Mountain Power is now the GMP Cow Power Program. Customers can elect to pay a premium on their electric bill, and that premium is passed directly to the farms in the program. In Sheldon, Vermont, Green Mountain Dairy has provided renewable energy as part of the Cow Power program. It started when the brothers who own the farm, Bill and Brian Rowell, wanted to address some of the manure management challenges faced by dairy farms, including manure odor, and nutrient availability for the crops they need to grow to feed the animals. They installed an anaerobic digester to process the cow and milking center waste from their 950 cows to produce renewable energy, a bedding to replace sawdust, and a plant-friendly fertilizer. The energy and environmental attributes are sold to the GMP Cow Power program. On average, the system run by the Rowells produces enough electricity to power 300 to 350 other homes. The generator capacity is about 300 kilowatts.

In Hereford, Texas, cow manure is being used to power an ethanol power plant. By switching to methane biogas, the ethanol power plant has saved 1000 barrels of oil a day. Over all, the power plant has reduced transportation costs and will be opening many more jobs for future power plants that will rely on biogas.

In Oakley, Kansas, an ethanol plant considered to be one of the largest biogas facilities in North America is using Integrated Manure Utilization System "IMUS" to produce heat for its boilers by utilizing feedlot manure, municipal organics and ethanol plant waste. At full capacity the plant is expected to replace 90% of the fossil fuel used in the manufacturing process of ethanol.

Europe

The level of development varies greatly in Europe. While countries such as Germany, Austria and Sweden are fairly advanced in their use of biogas, there is a vast potential for this renewable energy source in the rest of the continent, especially in Eastern Europe. Different legal frameworks, education schemes and the availability of technology are among the prime reasons behind this untapped potential. Another challenge for the further progression of biogas has been negative public perception.

In February 2009, the European Biogas Association (EBA) was founded in Brussels as a non-profit organisation to promote the deployment of sustainable biogas production and use in Europe. EBA's strategy defines three priorities: establish biogas as an important part of Europe's energy mix, promote source separation of household waste to increase the gas potential, and support the production of biomethane as vehicle fuel. In July 2013, it had 60 members from 24 countries across Europe.

UK

As of September 2013, there are about 130 non-sewage biogas plants in the UK. Most are on-farm, and some larger facilities exist off-farm, which are taking food and consumer wastes.

On 5 October 2010, biogas was injected into the UK gas grid for the first time. Sewage from over 30,000 Oxfordshire homes is sent to Didcot sewage treatment works, where it is treated in an anaerobic digestor to produce biogas, which is then cleaned to provide gas for approximately 200 homes.

In 2015 the Green-Energy company Ecotricity announced their plans to build three grid-injecting digesters.

Germany

Germany is Europe's biggest biogas producer and the market leader in biogas technology. In 2010 there were 5,905 biogas plants operating throughout the country: Lower Saxony, Bavaria, and the eastern federal states are the main regions. Most of these plants are employed as power plants. Usually the biogas plants are directly connected with a CHP which produces electric power by burning the bio methane. The electrical power is then fed into the public power grid. In 2010, the total installed electrical capacity of these power plants was 2,291 MW. The electricity supply was approximately 12.8 TWh, which is 12.6% of the total generated renewable electricity.

Biogas in Germany is primarily extracted by the co-fermentation of energy crops (called 'NawaRo', an abbreviation of *nachwachsende Rohstoffe*, German for renewable resources) mixed with manure. The main crop used is corn. Organic waste and industrial and agricultural residues such as waste from the food industry are also used for biogas generation. In this respect, biogas production in Germany differs significantly from the UK, where biogas generated from landfill sites is most common.

Biogas production in Germany has developed rapidly over the last 20 years. The main reason is the legally created frameworks. Government support of renewable energy started in 1991 with the Electricity Feed-in Act (*StrEG*). This law guaranteed the producers of energy from renewable sources the feed into the public power grid, thus the power companies were forced to take all produced energy from independent private producers of green energy. In 2000 the Electricity Feed-in Act was replaced by the Renewable Energy Sources Act (*EEG*). This law even guaranteed a fixed compensation for the produced electric power over 20 years. The amount of around 8 ¢/kWh gave farmers the opportunity to become energy suppliers and gain a further source of income.

The German agricultural biogas production was given a further push in 2004 by implementing the so-called NawaRo-Bonus. This is a special payment given for the use of renewable resources, that is, energy crops. In 2007 the German government stressed its intention to invest further effort and support in improving the renewable energy supply to provide an answer on growing climate challenges and increasing oil prices by the 'Integrated Climate and Energy Programme'.

This continual trend of renewable energy promotion induces a number of challenges facing the management and organisation of renewable energy supply that has also several impacts on the biogas production. The first challenge to be noticed is the high area-consuming of the biogas electric power supply. In 2011 energy crops for biogas production consumed an area of circa 800,000 ha in Germany. This high demand of agricultural areas generates new competitions with the food industries that did not exist hitherto. Moreover, new industries and markets were created in predominately rural regions entailing different new players with an economic, political and civil background. Their influence and acting has to be governed to gain all advantages this new source of energy is offering. Finally biogas will furthermore play an important role in the German renewable energy supply if good governance is focused.

Indian Subcontinent

Biogas in India has been traditionally based on dairy manure as feed stock and these "gobar" gas plants have been in operation for a long period of time, especially in rural India. In the last 2-3 decades, research organisations with a focus on rural energy security have enhanced the design of the systems resulting in newer efficient low cost designs such as the Deenabandhu model.

The Deenabandhu Model is a new biogas-production model popular in India. (*Deenabandhu* means "friend of the helpless.") The unit usually has a capacity of 2 to 3 cubic metres. It is constructed using bricks or by a ferrocement mixture. In India, the brick model costs slightly more than the ferrocement model; however, India's Ministry of New and Renewable Energy offers some subsidy per model constructed.

LPG (Liquefied Petroleum Gas) is a key source of cooking fuel in urban India and its prices have been increasing along with the global fuel prices. Also the heavy subsidies provided by the successive governments in promoting LPG as a domestic cooking fuel has become a financial burden renewing the focus on biogas as a cooking fuel alternative in urban establishments. This has led to the development of prefabricated digester for modular deployments as compared to RCC and cement structures which take a longer duration to construct. Renewed focus on process technology like the Biourja process model has enhanced the stature of medium and large scale anaerobic digester in India as a potential alternative to LPG as primary cooking fuel.

In India, Nepal, Pakistan and Bangladesh biogas produced from the anaerobic digestion of manure in small-scale digestion facilities is called gobar gas; it is estimated that such facilities exist in over 2 million households in India, 50,000 in Bangladesh and thousands in Pakistan, particularly North Punjab, due to the thriving population of livestock. The digester is an airtight circular pit made of concrete with a pipe connection. The manure is directed to the pit, usually straight from the cattle shed. The pit is filled with a required quantity of wastewater. The gas pipe is connected to the kitchen fireplace through control valves. The combustion of this biogas has very little odour or smoke. Owing to simplicity in implementation and use of cheap raw materials in villages, it is one of the most environmentally sound energy sources for rural needs. One type of these system is the Sintex Digester. Some designs use vermiculture to further enhance the slurry produced by the biogas plant for use as compost.

To create awareness and associate the people interested in biogas, the Indian Biogas Association was formed. It aspires to be a unique blend of nationwide operators, manufacturers and planners of biogas plants, and representatives from science and research. The association was founded in 2010 and is now ready to start mushrooming. Its motto is "propagating Biogas in a sustainable way".

In Pakistan, the Rural Support Programmes Network is running the Pakistan Domestic Biogas Programme which has installed 5,360 biogas plants and has trained in excess of 200 masons on the technology and aims to develop the Biogas Sector in Pakistan.

In Nepal, the government provides subsidies to build biogas plant at home.

China

The Chinese have experimented with the applications of biogas since 1958. Around 1970, China had installed 6,000,000 digesters in an effort to make agriculture more efficient. During the last years the technology has met high growth rates. This seems to be the earliest developments in generating biogas from agricultural waste.

In Developing Nations

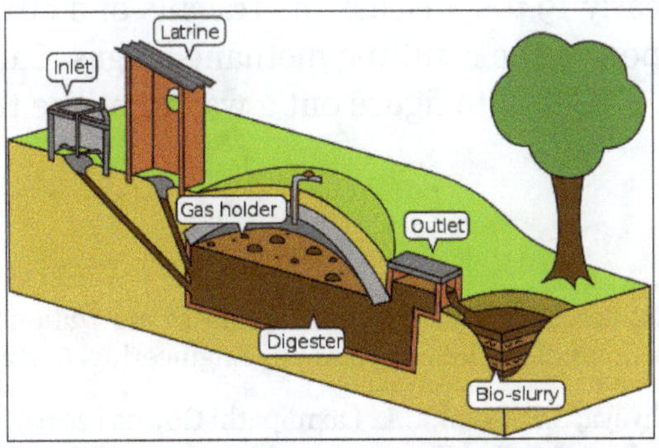

Simple sketch of household biogas plant.

Domestic biogas plants convert livestock manure and night soil into biogas and slurry, the fermented manure. This technology is feasible for small-holders with livestock producing 50 kg manure per day, an equivalent of about 6 pigs or 3 cows. This manure has to be collectable to mix it with water and feed it into the plant. Toilets can be connected. Another precondition is the temperature that affects the fermentation process. With an optimum at 36 C° the technology especially applies for those living in a (sub) tropical climate. This makes the technology for small holders in developing countries often suitable.

Depending on size and location, a typical brick made fixed dome biogas plant can be installed at the yard of a rural household with the investment between US$300 to $500 in Asian countries and up to $1400 in the African context. A high quality biogas plant needs minimum maintenance costs and can produce gas for at least 15–20 years without major problems and re-investments. For the user, biogas provides clean cooking energy, reduces indoor air pollution, and reduces the time needed for traditional biomass collection, especially for women and children. The slurry is a clean organic fertilizer that potentially increases agricultural productivity.

Domestic biogas technology is a proven and established technology in many parts of the world, especially Asia. Several countries in this region have embarked on large-scale programmes on domestic biogas, such as China and India.

The Netherlands Development Organisation, SNV, supports national programmes on domestic biogas that aim to establish commercial-viable domestic biogas sectors in which local companies market, install and service biogas plants for households. In Asia, SNV is working in Nepal, Vietnam, Bangladesh, Bhutan, Cambodia, Lao PDR, Pakistan and Indonesia, and in Africa; Rwanda, Senegal, Burkina Faso, Ethiopia, Tanzania, Uganda, Kenya, Benin and Cameroon.

In South Africa a prebuilt Biogas system is manufactured and sold. One key feature is that installation requires less skill and is quicker to install as the digester tank is premade plastic.

Society and Culture

In the 1985 Australian film *Mad Max Beyond Thunderdome* the post-apocalyptic settlement Barter town is powered by a central biogas system based upon a piggery. As well as providing electricity, methane is used to power Barter's vehicles.

"Cow Town", written in the early 1940s, discuss the travails of a city vastly built on cow manure and the hardships brought upon by the resulting methane biogas. Carter McCormick, an engineer from a town outside the city, is sent in to figure out a way to utilize this gas to help power, rather than suffocate, the city.

References

- Briens, Cedric; Piskorz, Jan; Berruti, Franco (2008). "Biomass Valorization for Fuel and Chemicals Production -- A Review". International Journal of Chemical Reactor Engineering. 6: R2. doi:10.2202/1542-6580.1674
- B.N. Divakara; H.D. Upadhyaya; S.P. Wani; C.L. Laxmipathi Gowda (2010). "Biology and genetic improvement of Jatropha curcas L.: A review". Applied Energy. 87 (3): 732–742. doi:10.1016/j.apenergy.2009.07.013
- R. E. Teixeira (2012). "Energy-efficient extraction of fuel and chemical feedstocks from algae". Green Chemistry. 14 (2): 419–427. doi:10.1039/C2GC16225C
- "Valcent Products Inc. Develops "Clean Green" Vertical Bio-Reactor". Valcent Products. Archived from the original on 18 June 2008. Retrieved 22, April 2020

- "Press Conference Launching International Biofuels Forum". United Nations Department of Public Information. 2007-03-02. Retrieved 13, August 2020
- The National Academies Press (2008). "Water Issues of Biofuel Production Plants". The National Academies Press. Retrieved 13, July 2020
- R. E. Teixeira (2012). "Energy-efficient extraction of fuel and chemical feedstocks from algae". Green Chemistry. 14 (2): 419–427. doi:10.1039/C2GC16225C
- Thomas F. McGowan, Michael L. Brown, William S. Bulpitt, James L. Walsh, Jr. (Editors) 2009. Biomass and Alternate Fuel Systems: An Engineering and Economic Guide. ISBN 978-0-470-41028-8 Wiley, 280 pages
- Hammerschlag, R (2006). "Ethanol's Energy Return on Investment: A Survey of the Literature 1999-Present". Environ. Sci. Technol. 40: 1744–50. doi:10.1021/es052024h

6

Concepts of Geothermal Energy

Geothermal energy is the thermal energy stored and generated inside the earth. Thermal energy refers to the energy that determines the temperature of the matter in the earth's crust. Geothermal energy is of two types – vapor-dominated form and liquid-dominated form. This chapter focuses on geothermal energy and its various applications.

Geothermal energy is the heat contained in the rock and fluid that fills the fractures and pores within the rock in the earth crust. The shallow ground water and hot rock in deep earth's crust to magma at interior of the earth are examples of geothermal energy resources.

Energy Inside the Earth

The temperature of the earth's core (at 4000 miles) with hot molten iron ranged from 2,800 to 6,000 degrees Celsius (°C). The geothermal energy (heat) continuously produced within the earth by slow radioactive decay of minerals which is natural in all rocks. The middle layer is mantle (1800 miles thick) which is partly rock and partly magma. The outer layer of the earth is crust with continents (15 to 35 miles thick) and ocean floors (3-5 miles thick).

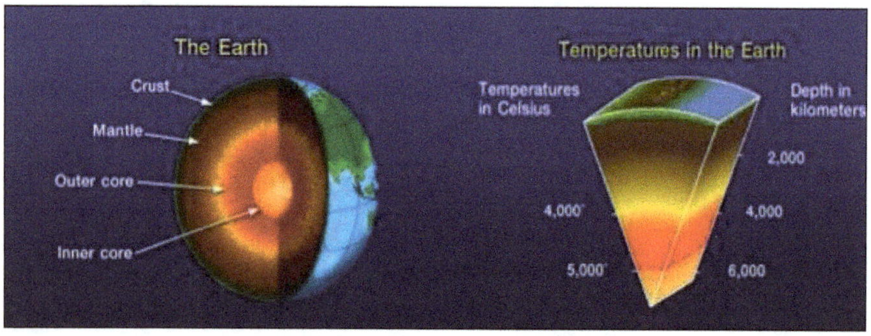

Temperature profile of earth layer wise.

The volcanoes are present in that part of crust which is thinned, faulted, or fractured by plate tectonics and magma come near the surface, on eruption lava moves out. The heat from magma was absorbed by rocks and water at depth. The heat transfer to water led to generation of geothermal energy. The temperature of the rocks and water increases with depth.

Formation of Geothermal Reservoir

Geothermal energy is called a renewable energy source because the water is replenished by rainfall, and the heat is continuously produced by the earth. The Earth's temperature increases with depth and it is higher than 4200 °C (7600 °F) at centre. The heat at the interior of earth is produced partly during earth's formation about 4.5 billion years ago and partly by continuous decay of radioactive isotopes. Due to plate tectonics, 12 huge plates are formed from the earth's crust and move apart or push together at a rate of millimeters per year. The collision of plates results in ocean trenches or strong earthquakes. The temperature at higher depth is capable of melting rock forming magma. The less dense hot magma reaches the earth's crust or as lava and carries heat energy.

The most portion of magma retains in earth crust and its high temperature heats the surrounding rocks and subterranean water. This hot water comes out to earth surface through faults and cracks as hot springs or geysers. The hot water and steam trapped in permeable rocks under a layer of impermeable rocks and is known as a geothermal reservoir. These reservoirs can act as sources of geothermal energy with potential to generate electricity or direct use application.

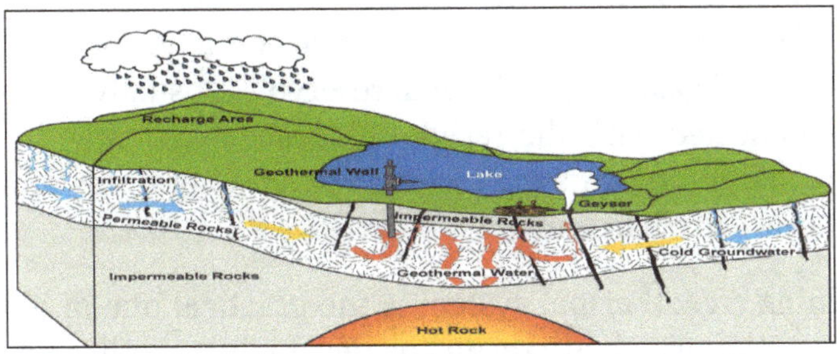

Formation of geothermal resource.

Classification of Geothermal Resources

The geothermal resources are classified based on enthalpy of the geothermal fluids that act as the carrier transporting heat from the deep hot rocks to the surface. Geothermal resources are divided into low, medium and high enthalpy resources.

Table: Classification of geothermal resources based on temperature.

Geothermal Resources	Temperature Range (average data)
Low enthalpy resources	<125 °C
Medium enthalpy resources	125°C - 225°C
High enthalpy resources	> 225°C

The temperature varies with depth for geothermal resources.

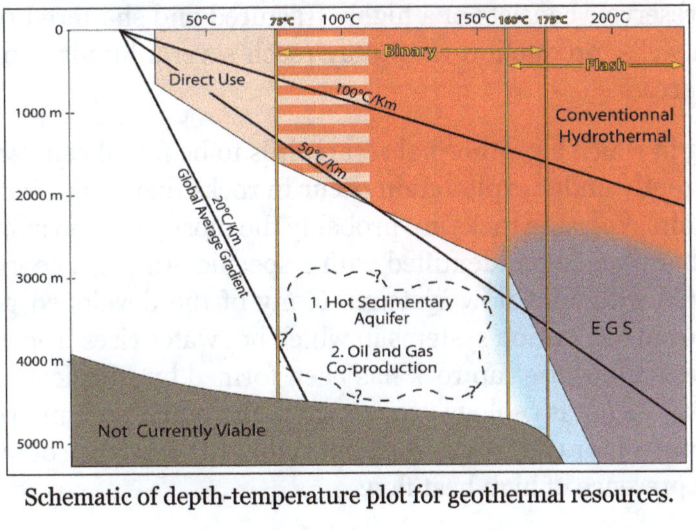

Schematic of depth-temperature plot for geothermal resources.

Geophysical Methods in Geothermal Exploration

Geophysical methods play a key role in geothermal exploration since many objectives of geothermal exploration can be achieved by these methods. The geophysical surveys are directed at obtaining indirectly, from the surface or from shallow depth, the physical parameters of the geothermal systems.

A geothermal system is made up of four main elements: a heat source, a reservoir, a fluid, which is the carrier that transfers the heat, and a recharge area. The heat source is generally a shallow magmatic body, usually cooling and often still partially molten. The volume of rocks from which heat can be extracted is called the geothermal reservoir, which contains hot fluids, a summary term describing hot water, vapour and gases. A geothermal reservoir is usually surrounded by colder rocks that are hydraulically connected with the reservoir. Hence water may move from colder rocks outside the reservoir (recharge) towards the reservoir, where hot fluids move under the influence of buoyancy forces towards a discharge area.

The first aspect of defining a geothermal system is the practical one of how much power can be produced. In most cases, the principal reason for developing geothermal energy is to produce electric power, although as an alternative the geothermal heat could be used in process applications or space heating. The typical geothermal system used for electric power generation must yield approximately 10 kg of steam to produce one unit (kWh) of electricity. Production of large quantities of electricity, at rates of hundreds of megawatts, requires the production of great volumes of fluid. Thus, one aspect of a geothermal system is that it must contain great volumes of fluid at high temperatures or a reservoir that can be recharged with fluids that are heated by contact with the rock. A geothermal reservoir should lie at depths that can be reached by drilling. It is unreasonable to expect to find a hidden geothermal reservoir at depths of less than 1 km; at the present time it is not feasible to search for geothermal reservoirs that lie at depths of more than 3 or 4 km.

Experience has shown that each well drilled in a geothermal field must be capable of supporting 5 MW of electrical production; this corresponds to a steam production of 10 tonnes/h. To accomplish this, a well must penetrate permeable zones, usually fractures that can support a high rate of flow. In many geothermal fields the wells are spaced so as to produce 25 – 30 MW km-2. At a few locations, where the reservoir consists of a highly fractured and shattered rock and there is little interference between wells, production rates may reach several hundred megawatts per square kilometre over small areas.

The geological setting in which a geothermal reservoir is to be found can vary widely. The largest geothermal fields currently under exploitation occur in rocks that range from limestone to shale, volcanic rock and granite. Volcanic rocks are probably the most common single rock type in which reservoirs occur. Rather than being identified with a specific lithology, geothermal reservoirs are more closely associated with heat flow systems. Many of the developed geothermal reservoirs around the world occur in convection systems in which hot water rises from deep within the earth and is trapped in reservoirs whose cap rock has been formed by silicification, or precipitation of other mineral elements. As far as geology is concerned, therefore, the important factors in identifying a geothermaı reservoir are not rock units, but rather the existence of tectonic elements such as fracturing, and the presence of high heat flow.

The high heat flow conditions that give rise to geothermal systems commonly occur in rift zones, subduction zones and mantle plumes, where, for unknown reasons, large quantities of heat are transported from the mantle to the crust of the earth.

Geothermal energy can also occur in areas where thick blankets of thermally insulating sediment cover basement rock that has a relatively normal heat flow. Geothermal systems based on the thermal blanket model are generally of lower grade than those of volcanic origin.

An important element in any model of a geothermal system in a volcanic area is the source for the system: in other words, the existence of an intruded hot rock mass beneath the area where the shallower reservoirs are expected to occur. The source of a geothermal system could be either a pluton or a complex of dykes, depending upon the rock type injected.

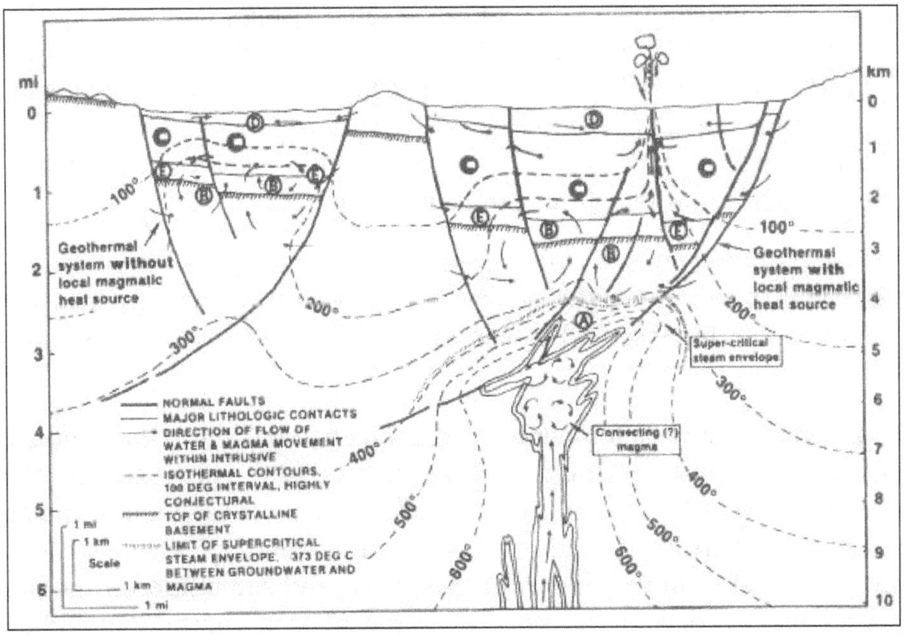

Diagrammatic cross sections of hypothetical geothermal systems. The system on the left is not closely associated to an intrusion but results from higher than normal thermal gradients through a sequence of thermally resistant rocks. The system on the right has an intrusion as the heat source.

A geothermal exploration programme is based on a number of phenomena associated with the intrusive model for a geothermal system. The model system that we hypothesise needs a source, in the form of intrusions that have been emplaced no more than half a million to one million years ago, so that excess heat still exists. The region above the source of the geothermal system must be fractured by tectonic activity, fluids must be available for circulation in a convection cell, and the precipitation of a cap rock must have taken place. All these elements represent targets for the application of geological, geophysical and geochemical prospecting techniques. Because of the high temperatures involved, both in the geothermal reservoir and in the source of the geothermal system, we can expect major changes to have taken place in the physical, chemical and geological characteristics of the rock, all of which can be used in the exploration project.

Heat is not easily confined in small volumes of rock. Rather, heat diffuses readily, and a large volume of a rock around a geothermal system will have its properties altered. The rock volume in which anomalies in properties are to be expected will, therefore, generally be large. Exploration techniques need not offer a high level of resolution. Indeed, in geothermal exploration we prefer an approach that is capable of providing a high level of confidence that geothermal fluids will be recovered on drilling.

A geothermal assessment programme on a regional basis will begin with a review and coordination of the existing data. All heat flow data acquired previously will have to be re- evaluated, re-gridded, smoothed, averaged and plotted out in a variety of forms in an attempt to identify areas with higher than normal average heat flow. Similarly, the volumes of volcanic products with ages younger than 10^6 years should be tabulated in a similar way to provide a longer-range estimate of anomalous heat flow from the crust. Because fracturing is important, levels of seismicity should be analysed, averaged and presented in a uniform format. All information on thermal springs and warm springs should be quantified in some form and plotted in the same format. Comparison of these four sets of data, which relate directly to the characteristics of the basic geothermal model described above, will produce a pattern that will indicate whether the area possesses the conditions favourable for the occurrence of specific geothermal reservoirs. These areas should then be tested further by applying some or all of the many geophysical, geological and geochemical techniques designed to locate specific reservoirs from which fluids can be produced.

The next stage of exploration consists of techniques for detecting the presence of a geothermal system. A geothermal system generally causes inhomogeneities in the physical properties of the subsurface, which can be observed to varying degrees as anomalies measurable from the surface.

These physical parameters include temperature (thermal survey), electrical conductivity (electrical and electromagetic methods), elastic properties influencing the propagation velocity of elastic waves (seismic survey), density (gravity survey) and magnetic susceptibility (magnetic survey). Most of these methods can provide valuable information on the shape, size, and depth of the deep geological structures constituting a geothermal reservoir, and sometimes of the heat source. An increase of silica amount in volcanic rocks decreases the density. Thermal surveys can delimit the areas of enhanced thermal gradient, which is a basic requirement for high-enthalpy geothermal systems, and define temperature distribution. Information on the existence of geothermal fluids in the geological structures can be obtained with the electrical and electromagnetic prospectings, which are more sensitive than other surveys to the presence of these fluids, especially if salty, and to variations in temperature. Moreover resistivity is strongly affected by porosity and saturation. Resistivity decreases with increasing porosity and increasing saturation. The same parameters measured indirectly from the surface can also be obtained from wells, by the method known as geophysical logging.

The depth at which rocks become conductive because of thermal excitation can be determined with relatively good reliability by means of the magnetotelluric method. Experience around the world has shown a remarkably good correiation between the depth to a thermally excited conductor and regional heat flow, as indicated in Fig. If thermally excited rocks occur at depths as shallow as 10 to 20 km within the crust, it is almost certain that a partially molten intrusive is present; the normal depth for thermally excited conductive rocks ranges from fifty to several hundred kilometres.

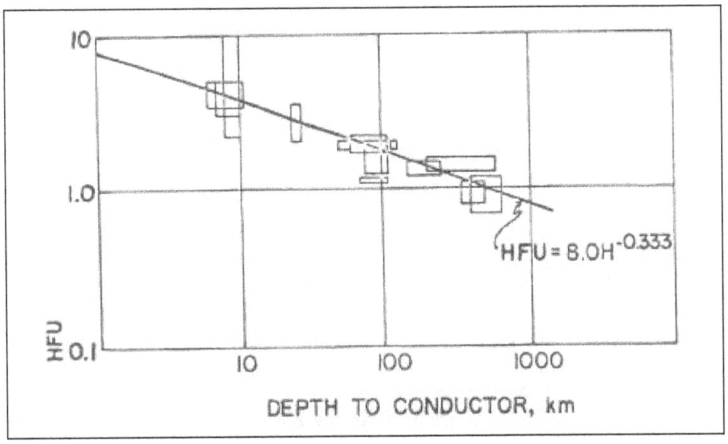

Observed correlation between the depth to thermally excited conductive zone in the crust or mantle (based on magnetotelluric soundings) and heat flow.

The Curie point method has the potential for providing confirmation of the existence of a hot rock mass in the crust. When rocks are heated above temperatures of a few hundred degrees Centigrade, they lose their ferromagnetism. Under favourable circumstances, the depth to this demagnetisation level can be determined with reasonable accuracy.

Further confirmation can be obtained by p-wave delay and shear wave shadow studies. (When an anomalously hot mass of rock is present in the ground, the compressional (p) waves from earthquakes are delayed in transit, while the shear (s) waves are reduced in amplitude). To detect such an effect, an array of seismograph stations is set up in the vicinity of an anomaly. The seismograph stations are operated over a sufficiently long time to record a few tens of teleseisms. The wave speeds for various ray paths through the suspected anomalous zone are then computed; if the rock is partially molten, the p-wave velocities will be from 20 to 30 per cent lower than their normal values. An increase of silica amount in volcanic rocks decreases the p-wave velocity. Moreover seismic velocity is strongly affected by porosity and saturation. Wave velocity is reduced by increasing porosity but shows different behaviour for different saturation, with an inverse relationship when saturation is high (100/85%) and a direct relationship when saturation is low, being constant for saturation of 15-85%.

A group of prospect areas should be defined, with reference to regional data and reconnaisance surveys. These areas may range in size from a few hundred to a thousand km². In rare cases, such as extensive thermal systems, they may be even larger. With the lack of resolution characteristic of the reconnaissance studies, it is unlikely that a prospect can be localised to an area of less than 100 km². Detailed geophysical, geological and geochemical studies will be needed in order to identify drilling locations once a prospect area has been defined from reconnaissance.

The objective of the more detailed studies is to identify the existence of a productive reservoir at attractive temperatures and depths. Geochemical surveys provide the most reliable indications of reservoir temperatures if the thermal fluids emerge at the surface. In any event, all springs and other sources of groundwater should be sampled and various geothermometer calculations carried out. Some prospect areas will probably show much more positive geochemical indicators than others. This could merely reflect the difference in the amount of leakage from subsurface reservoirs, but it does provide a basis for setting priorities for further testing; the geothermal reservoirs that show the most positive indications from geochemical thermometry should be the ones that are investigated first by other geophysical techniques.

The sequence in which geophysical methods are applied depends to a considerable extent on the specific characteristics of each prospect. It is not wise to define a particular sequence of geophysical surveys as being applicable to all potential reservoirs. In some cases, for example where we expect to find a subsurface convection system, various types of electrical survey could be highly effective in delineating the boundaries of the convecting system. Where large clay masses are present in the prospect area, on the other hand, electrical resistivity surveys can be deceptive. The particular type of electrical resistivity survey used at this stage is a matter of personal preference. Schlumberger sounding, dipole – dipole surveys, dipole mapping surveys and electromagnetic soundings can all be used to good effect. To some extent the choice of method here depends upon accessibility. The dipole–dipole traversing method and the Schlumberger sounding method are much more demanding in terms of access across the surface. The dipole mapping method and the electromagnetic sounding method can be applied in much more rugged terrain.

The objective when carrying out electrical surveys is to outline an area of anomalously low resistivity associated with a subsurface geothermal reservoir. When such an area has been identified, it is still necessary to confirm that the resistivity anomaly is the result of temperature and to locate areas within the anomaly where fracture permeability is likely to be high. Confirmation of subsurface temperatures is best done at this stage by drilling one or more heat flow boreholes. These heat flow holes need be only a few hundred metres deep if the area is one in which surface groundwater circulation is minimal. However, in volcanic areas where groundwater circulation occurs down to great depths, reliable heat flow data can be obtained only by drilling to one or two kilometres depth, in which case the heat flow hole becomes a reservoir test hole.

The number of heat flow holes that need to be drilled in a given prospect can vary widely; a single highly positive heat flow hole may be adequate in some cases while in others we may need several tens of heat flow holes to present convincing evidence for the presence of a geothermal reservoir at greater depth.

Once the probable existence of a geothermal reservoir has been established by a combination of resistivity studies and heat flow determinations, it is advisable to search for zones of fracture permeability in the reservoir before selecting a site for a test hole. Microseismic surveys are a widely used tool for studying activity in fracture zones in a prospect area. Surveys may require many weeks of observation in a given area. The accuracy with which active faults can be located using microearthquakes is often not good enough for the control of drill holes, although in some cases it is adequate. A potentially valuable by-product of a microearthquake survey is the determination of Poisson's ratio and related rock properties along various transmission paths through the potential geothermal system. Poisson's ratio and attenuation of seismic waves can be strongly affected by fracturing. The identification of anomalous areas of Poisson's ratio and p-wave attenuation can provide encouraging evidence for high permeability zones in the reservoir.

The most effective technique for studying a potential reservoir before drilling takes place is the seismic reflection method. Where there is a bedded structure to the subsurface this technique can be applied to detect faults through the disruption to the continuity of the bedding. The seismic reflection technique is extremely expensive, and a survey over a geothermal prospect may cost a significant fraction of the cost of a test well, but the results obtained with the seismic reflection method are usually much more conclusive than the results obtained with any other geophysical technique.

All of these geophysical surveys, targetted at defining the main characteristics of the geothermal reservoir, can also be supplemented with other types of geophysical surveys that assist us in understanding the regional geology and the local geological structure in a geothermal prospect. The self-potential survey is useful for our understanding of groundwater movement in an area. The gravity survey can be used to study the depth of fill in intermontaine valleys, and to locate intrusive masses of rock. Magnetic surveys can be used to identify the boundaries to flows in volcanic areas. Once all these detailed geophysical surveys have been carried out, a convincing set of data should be available before the decision is taken to locate a drill hole. There must be evidence for heat, there must be evidence for permeability, and the conditions for drilling must be established. Once we have achieved these objectives, we can tackle the problem of whether to drill a deep test well or not.

It is surprisingly difficult to record a true bottomhole temperature during the course of drilling a well. Mud is circulated through the well and removes much of the excess temperature as drilling progresses. In a closely controlled drilling programme, the temperature and volume of mud supplied to the well and recovered from the drilling operation should be monitored closely. Differences in the temperature between the mud going in and the mud coming back to the surface can be used to estimate bottomhole temperatures in a crude fashion. With the development of a mathematical model for the loss of heat from the rock to the drilling mud, it is conceivable that an even more precise temperature estimate will be attainable. The best temperature estimates during drilling are obtained by lowering maximum-reading thermometers to wellbottom whenever the bit has to be removed from the well for some reason. Several thermometers should be inserted at the same time in case one breaks or provides a false reading. They should be lowered to wellbottom on a heavy weight so as to be positioned as close as possible to the undisturbed rock at the bottom face of the borehole.

Subsurface Temperature and Thermal Gradient Surveys

The most direct method for studying geothermal systems is that of subsurface temperature measurements. Measurements can be made in holes as shallow as a few metres, but the preference at the present time is to conduct temperature surveys in wells that are at least 100 m deep. Temperatures measured a short distance beneath the surface of the earth are strongly affected by cyclic changes in temperature on the surface of the earth. Variations resulting from the diurnal temperature cycle penetrate only a few tens of centimetres into soils. The annual temperature cycle can make a significant contribution to temperature changes at many metres depth. Long-term climatic changes in temperature can conceivably cause barely detectable temperature effects at depths of 100 m. In order to obtain subsurface temperatures that are representative of heat flow from the interior of the earth, the temperature gradients have either to be measured at a depth beyond which the contribution from surface temperature changes is insignificant, or in such a way that the surficial effects can be removed. For example, if temperatures are measured at the bottom of shallow holes over a period of one year, then the annual temperature can be averaged out. Alternatively, if measurements are made at depths beyond which the annual wave penetrated significantly, the normal heat flow from the interior of the earth will be detected in a matter of a few days. It is still a matter of debate as to which approach is more effective.

Our objective in the case of thermal gradient measurements in boreholes is twofold: first, to detect areas of unusually high temperature, and second, to determine quantitatively the component of heat flow along the direction of the borehole. Unusually high temperatures could indicate the presence of geothermal activity. More quantitative results are obtained when thermal gradients are converted to heat flow through the use of Fourier's equation:

$$grade\,T = \Phi / K$$

Or,

$$\Delta T / \Delta z = \Phi_z / K$$

Where, $\Delta T / \Delta z$ the vertical gradient in temperature, K is the thermal conductivity and Fz is the thermal flux in the z direction. Thermal conductivity is measured in lab over rock samples, whereas temperature is measured in the well. The advantage of converting temperature gradients to heat flow values is that we eliminate the dependence on the thermal conductivity of the rock type. Thus, minor differences in temperature over a series of prospect holes can have added significance if we know that the differences are due not to a change in rock type, but to a change in the total amount of heat being supplied from beneath.

Determination of temperature in a test hole is not as easy as it might seem. In deep test holes, which must be drilled with a circulating fluid such as mud, a considerable disturbance of the normal temperature environment will take place during drilling. This is particularly true if the gradient is relatively high and the temperature change over the well interval is relatively large. As a rule of thumb, one must wait a period of time comparable to that involved in drilling the well before the well temperatures return to within 10 per cent of their undisturbed state. Drilling a well to several hundred metres depth may take a few days. In order to measure temperatures accurately for thermal gradient purposes we have to record downhole temperatures for periods that are several times longer than the duration of drilling. Fortunately there are a number of methods available for predicting stabilisation temperatures in a well. One such method consists of measuring temperatures at a given depth several times after completion of drilling. Temperatures are then plotted as a function of time on a linearising scale, defined as log $T/(T-s)$, where T is the total time since the drill first opened the borehole at the depth where the temperature is being measured, and s is the time since circulation in the well was halted. When temperatures are plotted against this linearised time scale, they fall along a straight line, provided the assumptions of the solution of Fourier's equation are valid: that is, that the rate of drilling was linear and that the removal of heat by the drilling mud from the bottom of the hole was consistent. The final temperature can be estimated quite accurately by extrapolating the linear relationship to zero linearised time, which corresponds to the temperature after an infinitely long re- equilibration process. An example of the use of this method is shown in figure.

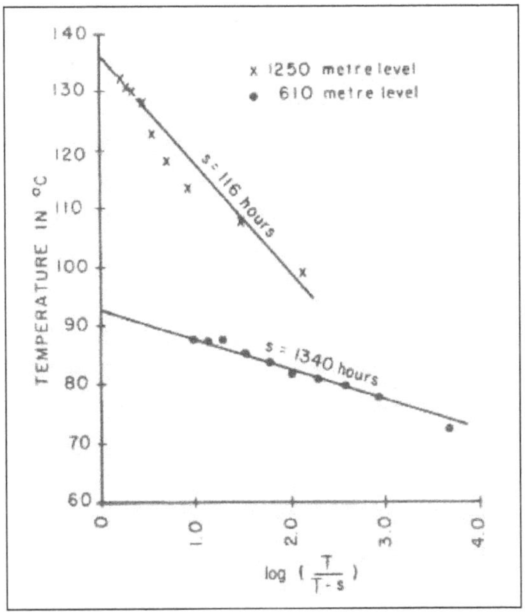

Extrapolation of borehole temperatures to equilibrium temperature at two depths in a borehole drilled at the summit of Kilauea volcano (Hawaii). The parameter s is the duration of circulation in the well following first penetration by drill bit, and T is the total time elapsed from first penetration to measurement of temperature.

After extrapolation to equilibrium temperature, other corrections must be applied to temperature gradient to determine the heat flow. These are the paleoclimatic correction, the topographic correction and the correction due to changes of ground surface due to rapid sedimentation or erosion. Topographic correction takes into account the temperature variation due to valleys and mountains. Indeed, isotherms follow the topography, raising below mountains and lowering below valleys. In many methods the external temperature is considered constant or a certain temperature distribution at surface is considered. The reference plane is then the horizontal one at the measuring site and crossing topography. This correction takes into account the shape of topography and its changes in the time, as long as isotherms are in equilibrium. However, when these changes are rapid equilibrium is not reached and corrections due to sedimentation or erosion must be considered.

Temperature variation at the surface due to climatic variations may affect variable ground depths depending on the wave length of variations. Climatic variations have different effects depending on their length and intensity. The recent variations affect mainly the first 200 m and the non-linearity of observed temperature is mostly due to them.

Paleoclimatic variations can be corrected using a schematic model where continuos climatic variations are taken into account and surface temperature is assumed constant in the last 10000 years.

The main difficulty in measuring heat flow in drill holes is that, in many geothermal fields, convection as well as conduction contributes significantly to total heat flow. Where convection is rapid, Fourier's simple equation cannot be used to compute heat flow. Convection is not easily computed since hydrogeological conditions are not well known. Heat transfer by convection can be computed by theoretical models. On the contrary, the extension and importance of circulation is defined comparing the observed situation with the theoretical model based on the assumption of simple conduction.

After all this corrections are applied to temperature, temperature gradient can be computed. To define the heat flow, however, thermal conductivity must be known. Another major problem in computing the heat flow is due to the variation of thermal conductivity, which can be measured many times and then averaged. Moreover conductivity decreases with temperature and increases with pressure. Variation due to pressure can be neglected since pressure changes very little at the depths involved in the investigation. Temperature variation cannot be neglected and depends also on the saturation level.

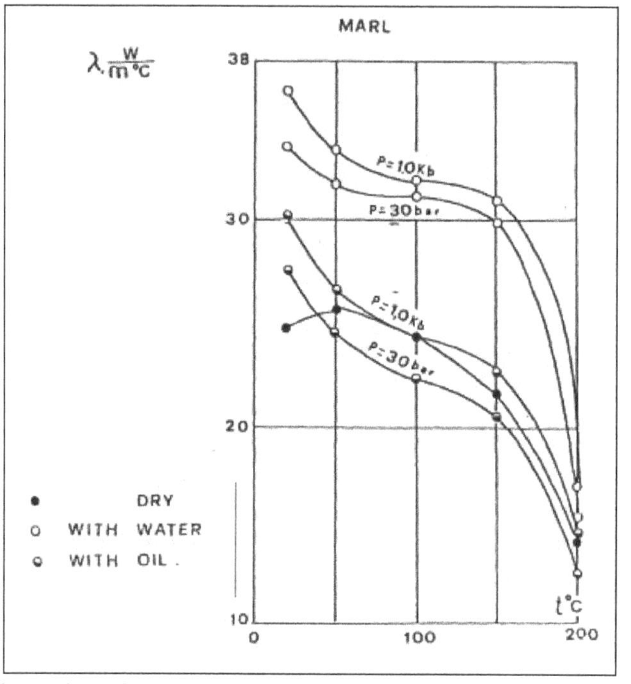

Thermal conductivity variation with temperature and pressure.

Usually conductivity increases with depth. This is probably due to compaction of rock at increasing pressures and to alignment of clay minerals and anisotropy effects. When thermal conductivity and temperature gradients are known, heat flow is mapped and anomalous areas are easily detected. Anomalies are linked to any heat source: however, those due to hot fluid circulation affect large areas. This way the most interesting areas can be defined and application of more expensive geophysical methods are limited.

Gravity Surveys

Gravity surveys are used during geothermal exploration to define lateral density variation related to deep magmatic body, which may represent the heat source. These anomalies can be created also by different degrees of differentiation of magma or variation in depth of crust-mantle interface which creates also depth variation of isotherms.

Gravity monitoring surveys are mainly performed in geothermal areas to define the change in groundwater level and for subsidence monitoring. Fluid extraction from the ground which is not rapidly replaced causes an increase of pore pressure and hence of density. This effect may arrive at surface and produce a subsidence, whose rate depends on the recharge rate of fluid in the extraction area and the rocks interested by compaction.

Repeated gravity monitoring associated to weather monitoring may define the relationship between gravity and precipitation which produces the shallow ground water level change. When gravity is corrected by this effect, gravity changes show how much of the water mass discharged to the atmosphere is replaced by natural inflow. The underground hydrological monitoring done by gravity survey is an important indication of the fluid recharge in geothermal systems and the need of reinjection.

Magnetic Surveys

Surveys of the spatial changes in the strength of the magnetic field over the surface of the earth have been used as a method for geophysical exploration for many years. Magnetite is the most common ferromagnetic mineral and so, in most cases, the magnetic permeability is controlled by the presence of varying amounts of magnetite and related minerals in the rock. The magnetic method has come into use for identifying and locating masses of igneous rocks that have relatively high concentrations of magnetite. Strongly magnetic rocks include basalt and gabbro, while rocks such as granite, granodiorite and rhyolite have only moderately high magnetic susceptibilities.

The magnetic method is useful in mapping near-surface volcanic rocks that are often of interest in geothermal exploration, but the greatest potential for the method lies in its ability to detect the depth at which the Curie temperature is reached. Ferromagnetic materials exhibit a phenomenon characterised by a loss of nearly all magnetic susceptibility at a critical temperature called the Curie temperature. Various ferromagnetic minerals have differing Curie temperatures, but the Curie temperature of titano-magnetite, the most common magnetic mineral in igneous rocks, is in the range of a few hundred to 570 °C. The ability to determine the depth to the Curie point would be an ability to determine the depth to the Curie point isotherm as well.

For magnetic field observations made at or above the surface of the earth, the magnetisation at the top of the magnetic part of the crust is characterised by relatively short spatial wavelengths, while the magnetic field from the demagnetisation at the Curie point in depth will be characterised by longer wavelength and lower amplitude magnetic anomalies. This difference in frequency characteristics between the magnetic effects from the top and bottom of the magnetised layer in the crust can be used to separate magnetic effects at the two depths and to determine the Curie point depth.

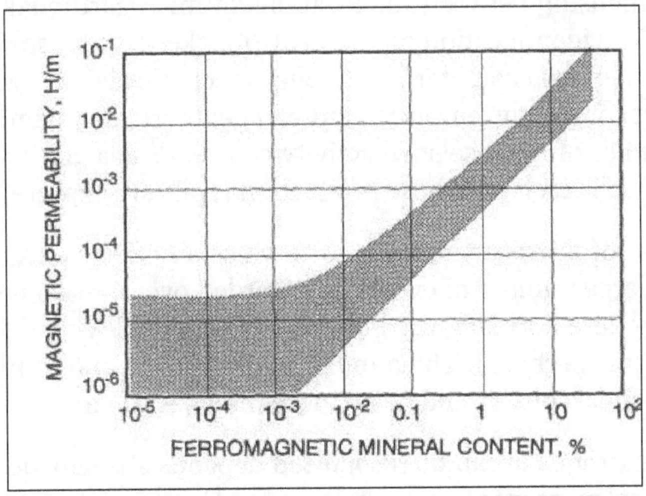

Correlation between magnetic permeability and the content of ferromagnetic minerals in instrusive rocks.

The magnetic field of the earth is of a complex nature because of its dipolarity. The inducing magnetic field has a dip angle that varies from place to place over the surface of the earth, and this feature introduces a complexity into the patterns of anomalies that are recorded. This problem has long been recognised in the analysis of magnetic data, and a procedure has been developed to re-compute the magnetic profile map for a vertical inducing field using the actual observed magnetic map. This first step in processing magnetic data is termed the conversion of the magnetic map 'to the pole', or to the form it would have for a vertical inducing field.

Another difficulty when dealing with observed magnetic fields is that the specific character of the anomaly depends, in a complicated way, on the size and shape of the magnetic bodies. For simple bodies with well-defined tops and bottoms, such as a pluton with magnetisation disappearing at a specific depth, the magnetic anomaly can be computed numerically either in the space domain or the spatial frequency domain. The fact that a real body has lateral extent affects the spectrum that is computed, meaning that a set of curves must be computed for prismatic bodies that are reasonably representative of a given anomaly in order to determine the depth extent of magnetisation by comparison of the spectra computed from an observed magnetic map with the spectra computed for a prismatic body. When interpreting a magnetic map, the most straightforward procedure is to select a relatively simple magnetic anomaly on the map that can be represented by a body of simple geometric shape. A series of spectra is computed for various depths to the Curie point, and the spectra computed from the field data are compared with these to reach an estimate of the depth to the Curie point.

Seismic Methods

These methods can be divided into two main subclasses: passive seismic methods, which deal with the effects of natural earthquakes or those induced by fracturing related to geothermal fluid extraction and injection; and active seismic methods, which cover all seismic prospectings having an artificial wave source.

Passive Seismic Methods

It has been observed by many researchers that geothermal systems occur mainly in areas characterised by a relatively high level of microseismic activity. However, in detail there does not appear to be a one-to-one relationship between the locations of microearthquakes and the presence of geothermal reservoirs. The identification of microearthquakes in a prospect area serves primarily as a means of investigating and characterising modern tectonic activity, which may be controlled by the same factors that control the emplacement of a geothermal system; in particularly favourable circumstances, studies of microseismic activity can serve as a guide when drilling into fractured rocks in a geothermal reservoir whose production levels are expected to be high.

In order for the location of microearthquakes to serve as an effective exploration tool, it is necessary that a relatively large number of events be recorded over a reasonable recording period in a survey area. The procedure normally used at present is to deploy a number of highly sensitive seismograph units in a prospect area. The number may range from 6 to 12 or more. The distance between any pair of seismographs should be no more than 5 – 10 km.

The range to which an earthquake can be recognised depends strongly on the energy released by that earthquake. The usual measure of energy release used by seismologists is the Richter M factor, which is a logarithmic scale indicating the amplitudes of seismic waves. A magnitude 4 earthquake is normally felt by a few people within several kilometres of the epicentre. Such earthquakes occur relatively rarely, perhaps only a few times a year, even in areas with a high level of seismicity. However, almost all sets of earthquakes for which magnitudes have been measured follow an inverse linear relationship between the logarithm of the number of earthquakes and their magnitudes. A decrease in magnitude of one Richter unit is normally accompanied by an increase of a factor of 10 in the number of events that occur within a given time. This relationship is the rationale for using microearthquakes rather than larger size earthquakes to estimate seismicity in a prospect area. If the threshold of the field system for detecting a microearthquake can be improved to the point where events of magnitude – 1 or – 2 can be recognised, the number of events that will be recorded in a given time can be increased by several orders of magnitude. In such a case, over a recording interval of 30 to 60 days, one would reasonably expect to record hundreds of events in an area of seismic activity. In aseismic areas, on the other hand, few if any events would be recorded even at such a low threshold for detection.

When arrival times for a micro-earthquake have been recorded at several stations in an array, the problem remains of determining the location at which that earthquake occurred. In carrying out this calculation, it is usually assumed that the earthquake energy was released from a point source, and that the seismic waves travelled through a uniform earth to be recorded at each seismometer. This assumption is not obviously valid, but programs to determine epicentres when there are lateral variations in wave speed have not yet come into general use.

In locating an epicentre, the unknowns to be determined are the x, y, and z coordinates of the origin, the time at which the event occurred, and the subsurface wave speed distribution. Given a constant wave speed assumption, the location of an epicentre can be found from only four p arrival times. However, the equations are not always stable because the wave speed may not be the same for all travel paths. Likewise, if there are variations in velocity from point to point in the medium, the assumption of straight line travel from the origin to each receiver location is not valid and the solution may not work. An alternative procedure which may work better involves the use of s arrival times as well as p arrival times. In this case, a solution can be obtained using p and s arrivals recorded at three stations. A Wadati diagram, which is a cross-plot between the p arrival times and the s – p arrival-time differences, is constructed. If the trend of p versus p – s times is projected to zero p – s difference, the origin time for the earthquake is specified. Once an origin time is known, only three arrival times are needed to obtain a solution for the coordinates of an earthquake.

In addition to determining epicentres of local earthquakes in a geothermal prospect, information about the geology and tectonics can be obtained from fault plane solutions and first motion studies of these earthquakes. The directions of first motion for p waves can be analysed by plotting the sense of motion on an upper-hemisphere stereographic projection centred on the epicentre. In this diagram compressional first arrivals are denoted by solid circles and dilations by open circles. Generally, the sense of first motion forms a pattern on the stereonet projection that can be bounded by the traces of two planes. One of these planes is the fault plane itself, and the other is an auxiliary fault plane, that is, a plane perpendicular to the actual fault plane in whose pole is the pole of motion. Because two planes are needed to form a pattern that encloses the first arrivals, ambiguity as to which one is the actual fault plane exists unless other information is available to make the selection.

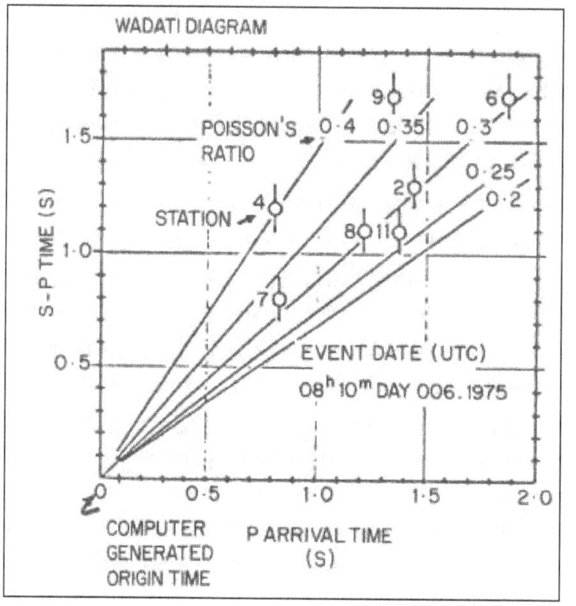

Example of a "Wadati plot" for a microseismic survey of the East Rift zone on the Island of Hawaii. Arrivals of p and s waves recorded on a close-spaced array of seven seismograph stations. The sloping lines are the loci of points with the same Poisson's ratio.

The use of fault plane solutions is valuable in determining whether the earthquake activity in a prospect area is anomalous or typical of the region. An example of such an application is shown in figure, where the fault plane solution observed for earthquakes within a small prospect area shown at the top of the map is compared with earlier more general fault plane solutions in the same area. The direction of tension as indicated by the fault plane solution is the same in the prospect area as in the region, indicating that the local earthquake distribution is controlled by the same crustal phenomena.

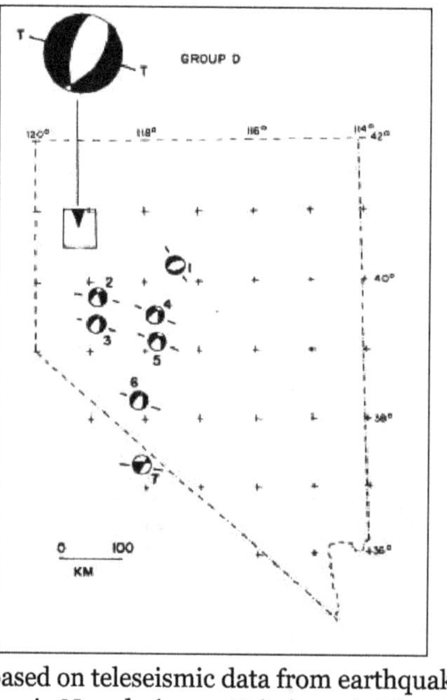

Comparison of first motion diagrams based on teleseismic data from earthquakes in Nevada and one based on local earthquake data from microseismic survey in Nevada (group D in insert). Directions of tensional stresses are uniform, indicating that events detected during the microseismic survey are probably not anomalous.

Another potentially valuable determination that can be made is that of Poisson's ratio. Poisson's ratio is the ratio of lateral strain to strain in the direction of stress when a cylinder of rock is subjected to uniaxial stress. Poisson's ratio is also definable in terms of the ratio of compressional to shear wave velocities. Poisson's ratio can be important in geothermal exploration inasmuch as it seems to be an indicator of the degree of fracturing in a rock. Both experimental and theoretical analyses have indicated that extensive fracturing of a liquid-filled rock causes Poisson's ratio to be higher than normal. This causes a minor reduction in the p-wave velocity and a significant reduction in s-wave velocity.

Poisson's ratio is determined from Wadati plots as described previously; that is, from plots of p – s arrival times as a function of p arrival times. Figure shows a presentation of Poisson's ratio determinations on a plane map. Each value is written along the assumed ray path from the epicentre to the receiver location; although there is no certainty that straight- line wave propagation took place. The significant anomaly in Poisson's ratio characterised in these data corresponds to the location of a successful geothermal well that was subsequently drilled.

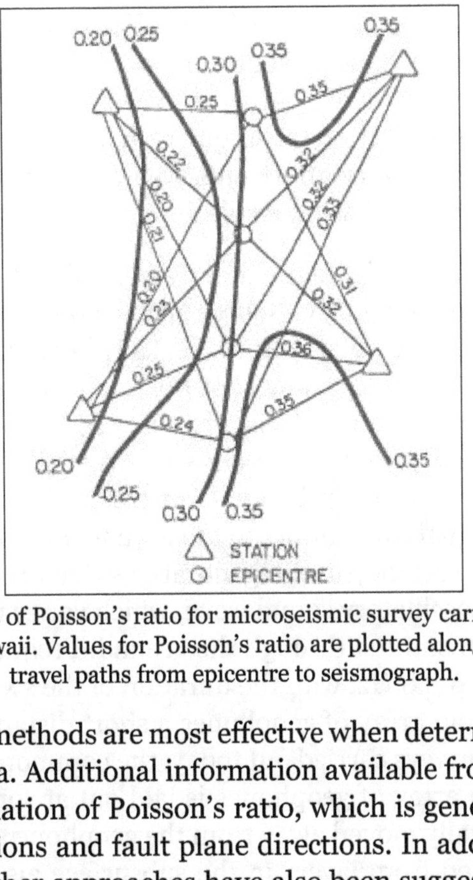

Results of determinations of Poisson's ratio for microseismic survey carried out on East Rift zone of Kilauea Volcano, Hawaii. Values for Poisson's ratio are plotted along apparent straight-line travel paths from epicentre to seismograph.

In summary, passive seismic methods are most effective when determining the epicentres of small earthquakes in a prospect area. Additional information available from such micro-earthquake locations includes the determination of Poisson's ratio, which is generally indicative of fracturing, and the direction of first motions and fault plane directions. In addition to these applications of passive seismology, several other approaches have also been suggested. These include the detection of p-wave delays and the observation of ground noise. If an increase in temperature results in the reduction of p-wave velocity over a large volume in the crust, the measurement of delay times from teleseisms, or distant earthquakes, could be used to locate large hot bodies that act as the source of geothermal systems. This technique has been the subject of considerable research in recent years, but has been faced by practical difficulties during its current deployment. Teleseisms that occur at great enough distances to be of use in p-wave delay surveys occur only rarely, perhaps a few times a month, with the magnitudes needed to obtain accurate p-wave arrival times. In order to obtain a detailed picture of subsurface structure, many hundreds of p-wave arrivals must be recorded. This can only be done by using a few stations over a long period of time or a very large number of stations for a shorter period of time. Each approach represents a relatively large investment.

Microseismicity is often not only natural but also induced by geothermal activity. One major cause appears to be injection, which results in the reservoir rock being rapidly cooled.At the Geysers geothermal field injection has increased by about 50% the number of M=2.4 events being recorded, with no increase observed in M=2.5 events.

Active Seismic Methods

Both seismic reflection and seismic refraction surveys have been used in geothermal exploration. Seismic refraction surveys have been used to a limited extent because of the amount of effort required to obtain refraction profiles giving information at depths of 5 to 10 km, and the problems caused by the generally high degree of complexity of geological structures in areas likely to host geothermal systems. On the other hand, standard seismic reflection surveys have often yielded surprisingly useful results, even in areas that were expected to prove difficult. The primary requirement for using the seismic reflection technique is that the subsurface be laminar in acoustic properties so that reflectors can be traced horizontally and interruptions in reflectors can be used to identify faults where displacement has taken place.

Seismic reflection surveys can be carried out either by exploding dynamite in shot holes as an energy source or by using some non-explosive source such as the Vibroseis. Considering that most geothermal prospect areas are made up of volcanic rocks either in the subsurface or near the subsurface, experience has indicated that drilling shot holes is probably not a practical approach to obtaining seismic reflection data. The Vibroseis technique seems to be more generally applicable in such difficult areas. A typical Vibroseis source system will use from one to five truck-mounted vibrators, usually having masses of the order of 15 tons. The vibrators are operated in unison to increase the intensity of the signal transmitted to the ground, and occasionally are spaced in such a way as to cancel some of the surface travelling vibrations that are a noise source at the receiver. In the Vibroseis approach, an oscillatory sound wave, which varies in frequency over the duration of a single transmission, is transmitted through the ground. Frequency is normally swept from a high of 60 – 80 Hz to a low of 6 – 8 Hz, with the duration of the sweep being 8 – 10 s. In order to obtain recognisable signals at an array of geophones a short distance away, many transmissions are stacked. In most seismic surveys carried out today, the common depth-point stacking array of geophones is used. In this, an array of geophones is laid out at some distance from the vibrator location. The vibrator is gradually moved away from the geophone spread so that the point where the acoustic waves are reflected from layers in the subsurface also moves away. The geophone spread is then moved a short distance along the profile and the process repeated. In analysing the ray paths, it can be seen that reflections are obtained from the same point in the subsurface with different separations between the vibrator and the geophones. Data quality can be enhanced by synchronously adding reflections that arrive from the same reflection point on a subsurface interface, but in so doing corrections must be made for the difference in travel times that is associated with the difference in separation from transmitter to receiver.

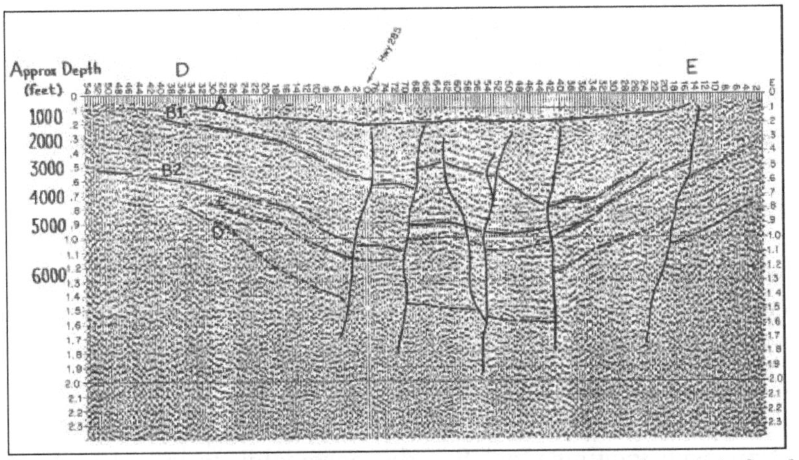

Seismic reflection profile in Colorado. Events B1 and B2 are volcanic flows, C is flow bottom, D is probably Paleozoic carbonate sequence. Near-vertical lines are faults.

It is often thought that good-quality seismic sections cannot be obtained in volcanic terrain. This is not always a serious problem; an example of a seismic section obtained in an area where basalt flows are interbedded with gravels and shale or alluvium is shown in figure (the area is in south-central Colorado, USA, near the mineral hot springs). The prominent reflectors seen on the seismic time section are the tops and bottoms of the basalt flows. They can be traced quite well, with faulting being readily apparent, along with a thinning-out of the basalt flows towards the right-hand side of the section. The value of the seismic method is shown by the cross-section in Figure. which was compiled from an interpretation of the seismic time section as shown in Figure. and other electrical data obtained with the Schlumberger sounding method. An area of relatively low resistivity, which is believed to represent the reservoir feeding the mineral hot springs, is seen to be associated with a down-drop fault block that is traced by the surface of the basalt flows.

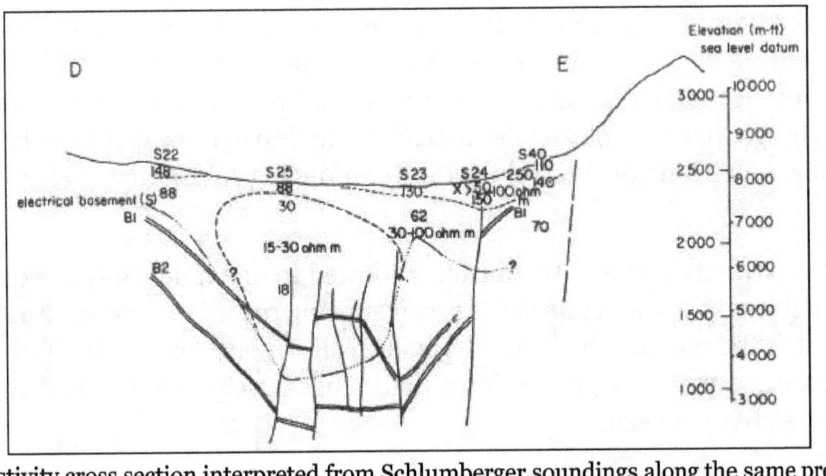

Resistivity cross section interpreted from Schlumberger soundings along the same profile as seismic section shown in figure reflective horizon from figure are shown.

Procedures for acquiring and processing seismic reflection data are well developed. Extensive computer facilities must be available to permit reduction of field observations to time sections such as are shown here, and to enhance the events that serve to trace the marker horizons in the section. An added merit of modern data processing as used in seismic reflection analysis is that, by varying the separation between vibrator and geophone spread, it is possible to determine the average acoustic wave speed to the depth of a reflector. When a series of reflectors is apparent on a seismic section, a seismic velocity profile can be extracted from the measurements. Seismic velocities themselves are useful in recognising anomalies caused by temperature in relatively shallow rocks.

Geophysical Well Logging and Bore Hole Methods

Logging in geothermal exploration suffers the main problems of high temperature and strong corrosion which can corrupt results (and tools!). It often requires a cooling of the well and the use of cables and instruments specifically built, hence increasing the prize. Logging is cheap compared to drilling, but is quite expensive by itself. However it is a valuable tool for direct exploration and detection of physical parameters which can then be extrapolated by surface geophysical methods. For direct exploration, well logging identifies the permeability, in particular fractures, and the nature of the fluids present. When possible, geophysical logs are recorded simultaneously. Most measurements can be made only where the hole has not been cased. Moreover, logs must be performed soon after wells are drilled otherwise rocks may change their log characteristics as a result of standing open to drilling fluids and the consequent penetration of the mud filtrate into the borehole. Resistivity, radio-active (gamma-ray), seismic wave velocity (acoustic and sonic) and density logs are the most common. Multi-frequency electromagnetic logging is becoming popular and when many wells are logged, the EM tomography method may define the distribution of resistivity around the borehole and the whole area interested by drilling.

Besides well logging, the use of drill holes to place the equipment in an advantageous position to explore subsurface can be an important part of an exploration program. Instrumentation must be of small diameter, water tight and designed to operate under adverse environmental conditions.

The mise a la masse method is one of the oldest methods used in-hole and is used when a prospect hole penetrates a highly conductive zone. A current electrode is embedded in the conductive zone and energized with direct current, the other electrode being a large distance away on surface. The extent, dip, strike and continuity of the zone will be better indicated by introducing the current directly into it than by the usual mapping techniques. Lately, methods using the casing pipe itself as a charged current electrode instead of the buried point source have been used to monitor fluid flow behaviour during reinjection and during hydraulic fracturing operations in hot dry rocks.

In the cross-hole resistivity survey a trasmitter is located in one hole while a receiver is drawn up another, with the strength of the signal at the receiver being recorded. The attenuation of the signal as it travels through the medium between the holes will depend on the resistivity structure. In a general sense, attenuation will be greater if the field senses a region of low resistivity, and lower if a region of high resistivity is seen.

Uses of Geothermal Energy

Geothermal energy use can be divided into three categories: direct-use applications, geothermal heat pumps (GHPs), and electric power generation.

Direct uses

Probably the most widely used set of applications involves the direct use of heated water from the ground without the need for any specialized equipment. All direct-use applications make use of low-temperature geothermal resources, which range between about 50 and 150 °C (122 and 302 °F). Such low-temperature geothermal water and steam have been used to warm single buildings, as well as whole districts where numerous buildings are heated from a central supply source. In addition, many swimming pools, balneological (therapeutic) facilities at spas, greenhouses, and aquaculture ponds around the world have been heated with geothermal resources. Other direct uses of geothermal energy include cooking, industrial applications (such as drying fruit, vegetables, and timber), milk pasteurization, and large-scale snow melting. For many of those activities, hot water is often used directly in the heating system, or it may be used in conjunction with a heat exchanger, which transfers heat when there are problematic minerals and gases such as hydrogen sulfide mixed in with the fluid.

Bagno Vignoni: Hot springs.

Geothermal Heat Pumps

Geothermal heat pumps (GHPs) take advantage of the relatively stable moderate temperature conditions that occur within the first 300 metres (1,000 feet) of the surface to heat buildings in the winter and cool them in the summer. In that part of the lithosphere, rocks and groundwater occur at temperatures between 5 and 30 °C (41 and 86 °F). At shallower depths, where most GHPs are found, such as within 6 metres (about 20 feet) of Earth's surface, the temperature of the ground maintains a near-constant temperature of 10 to 16 °C (50 to 60 °F). Consequently, that heat can be used to help warm buildings during the colder months of the year when the air temperature falls below that of the ground. Similarly, during the warmer months of the year, warm air can be drawn from a building and circulated underground, where it loses much of its heat and is returned.

A GHP system is made up of a heat exchanger (a loop of pipes buried in the ground) and a pump. The heat exchanger transfers heat energy between the ground and air at the surface by means of a fluid that circulates through the pipes; the fluid used is often water or a combination of water and antifreeze. During warmer months, heat from warm air is transferred to the heat exchanger and into the fluid. As it moves through the pipes, the heat is dispersed to the rocks, soil, and groundwater. The pump is reversed during the colder months. Heat energy stored in the relatively warm ground raises the temperature of the fluid. The fluid then transfers this energy to the heat pump, which warms the air inside the building.

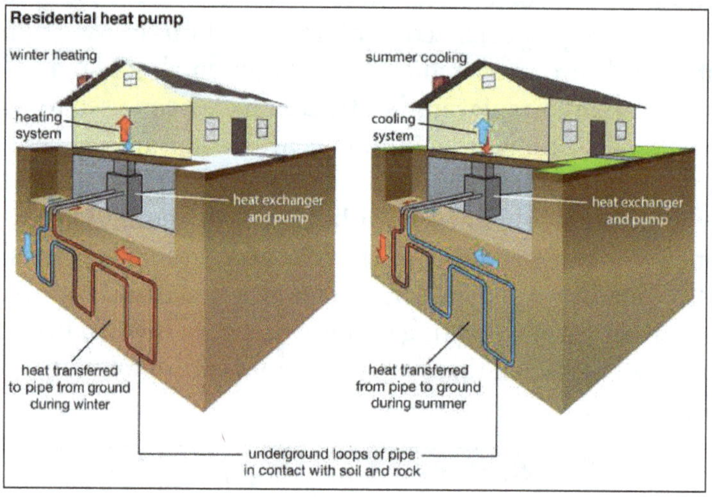

Residential heat pump: Residential heat pump operation for summer cooling and winter heating.

GHPs have several advantages over more conventional heating and air-conditioning systems. They are very efficient, using 25–50 percent less electricity than comparable conventional heating and cooling systems, and they produce less pollution. The reduction in energy use associated with GHPs can translate into as much as a 44 percent decrease in greenhouse gas emissions compared with air-source heat pumps (which transfer heat between indoor and outdoor air). In addition, when compared with electric resistance heating systems (which convert electricity to heat) coupled with standard air-conditioning systems, GHPs can produce up to 72 percent less greenhouse gas emissions.

Electric Power Generation

Depending upon the temperature and the fluid (steam) flow, geothermal energy can be used to generate electricity. Geothermal power plants can produce electricity in three ways. Despite their differences in design, all three control the behaviour of steam and use it to drive electrical generators. Given that the excess water vapour at the end of each process is condensed and returned to the ground, where it is reheated for later use, geothermal power is considered a form of renewable energy.

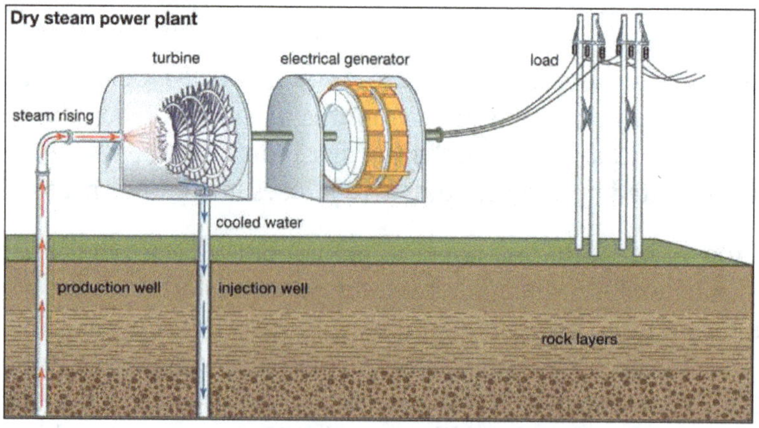

Dry steam geothermal power generation.

Some geothermal power plants simply collect rising steam from the ground. In such "dry steam" operations, the heated water vapour is funneled directly into a turbine that drives an electrical generator. Other power plants, built around the flash steam and binary cycle designs, use a mixture of steam and heated water ("wet steam") extracted from the ground to start the electrical generation process.

In flash steam power plants, pressurized high-temperature water is drawn from beneath the surface into containers at the surface, called flash tanks, where the sudden decrease in pressure causes the liquid water to "flash," or vaporize, into steam. The steam is then used to power the turbine-generator set. In contrast, binary-cycle power plants use steam driven off a secondary working fluid (such as ammonia and hydrocarbons) contained within a closed loop of pipes to power the turbine-generator set. In this process, geothermally heated water is drawn up through a different set of pipes, and much of the energy stored in the heated water is transferred to the working fluid through a heat exchanger. The working fluid then vaporizes. After the vapour from the working fluid passes through the turbine, it is recondensed and piped back to the heat exchanger.

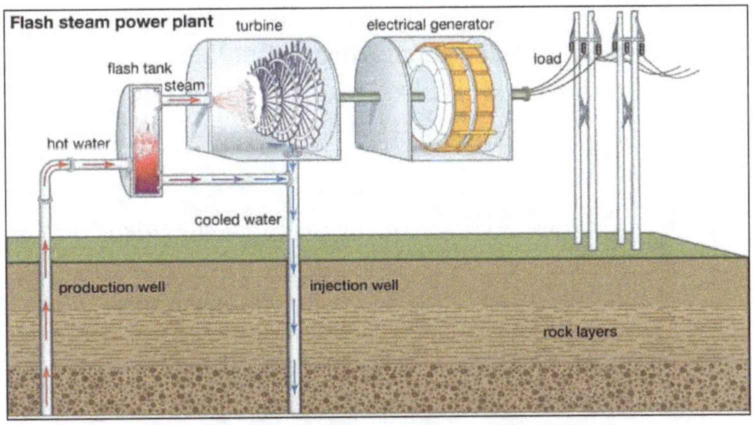

Flash steam geothermal power generation.

Electrical power usually requires water heated above 175 °C (347 °F) to be economical. In geothermal plants using the Organic Rankine Cycle (ORC), a special type of binary-cycle technology that utilizes lower-temperature heat sources (such as biomass combustion and industrial waste heat), water temperatures as low as 85–90 °C (185–194 °F) may be used.

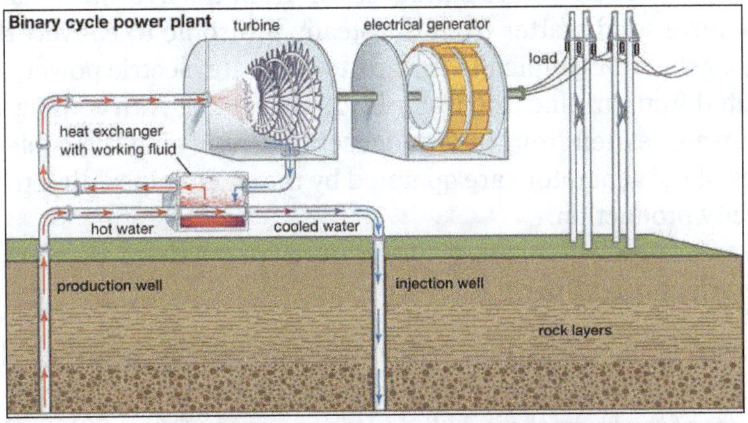

Binary-cycle geothermal power generation.

Types of Geothermal Power Plants

Geothermal power plants use hydrothermal resources for electricity generation from the heat source (magma) within the earth's crust. It consists of water (hydro) and heat (thermal). The high temperature (150 to 350 °C) in hydrothermal resources is harnessed either by drilling wells (dry steam wells or hot water wells) into the earth and piping the steam or hot water to the surface. The power plants use steam to power a turbine for generation of electricity. The geothermal power plants are categorized into three groups:

- Dry steam geothermal plant.
- Flash steam geothermal plants.
- Binary power plants.

Dry Steam Geothermal Plant

A dry steam geothermal plant is a type of geothermal power plant in which steam, generated underground by geothermal heat, is used directly, thus eliminating the need for boilers and boiler fuel that characterizes other steam-power-generating technologies. Conventional turbine generators are used with dry steam resources. It is also known as vapour dominated plant.

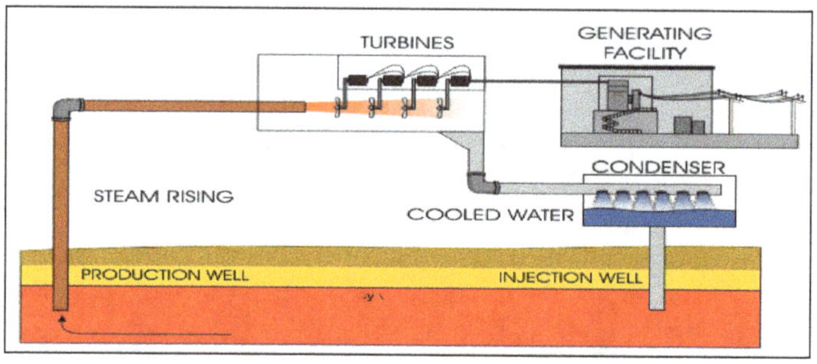

Schematic diagram of dry steam geothermal plant.

Water boils underground in a hydrothermal resource when it has pressure of about 7 atm and temperature of about 165 °C. The dry steam fields are located at the Geysers region of California and Matsukawa region of Japan.

The plant consists of production well to extract steam from the hydrothermal resource, a centrifuge separator to remove solid matter from the steam, a turbine to convert thermal energy into mechanical energy, a generator coupled to turbine to generate electric power, a condenser to condense wet steam exited from turbine into water by direct contact with cooling water and a cooling tower to cool warm water exited from the condenser and returning the cooled water to the condenser. The turbines of the generators are operated by the steam piped directly from a geothermal reservoir for electricity production.

Flash Steam Geothermal Plant

A flash-steam geothermal plant is a type of geothermal steam plant where liquid is flashed to steam and the steam is separated from the remaining liquid used to drive a turbine generator. When the temperature of the hydrothermal liquids is over 350 °F (177 °C), flash-steam technology is generally employed. It is also known as wet steam or liquid dominated geothermal plant. The dual-flash cycle which separates the steam at two different pressures return water to geothermal reservoir improves the economics of most hydrothermal flash plant. The dual-flash cycle produces 20% to 30% more power than a single-flash system at the same fluid flow.

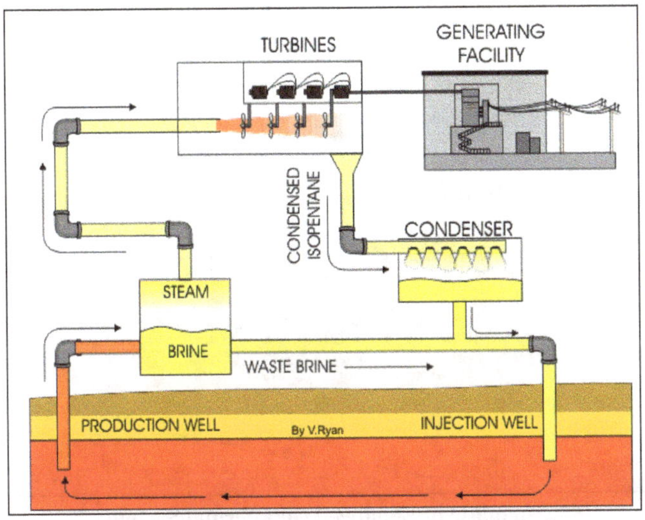

Schematic diagram of flash steam geothermal plant.

These plants harness high-pressure hot water from deep inside the earth and convert it to steam to drive the generator turbines. The water is condensed after cooling of steam and is injected back into the ground for recycling. Most of the geothermal power plants are flash steam plants.

Binary Cycle based Geothermal Plant

Binary cycle power plants are used with hot water that is at a lower temperature (100 – 300 °F). The hot water is passed through a heat exchanger in conjunction with a secondary fluid with a lower boiling point. Generally a hydrocarbon such as iso-butane or iso-pentane I used as secondary fluid. The secondary fluid vaporizes, which turns the turbines, which drive the generators. The remaining secondary fluid is simply recycled through the heat exchanger. The geothermal fluid is condensed and returned to the reservoir.

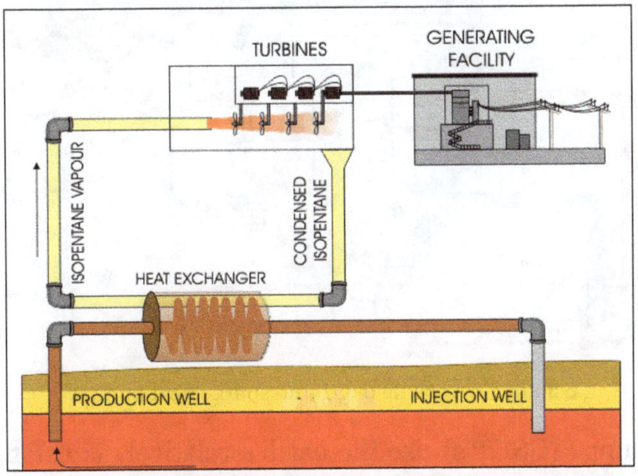

Schematic diagram of binary cycle based geothermal plant.

Enhanced Geothermal System

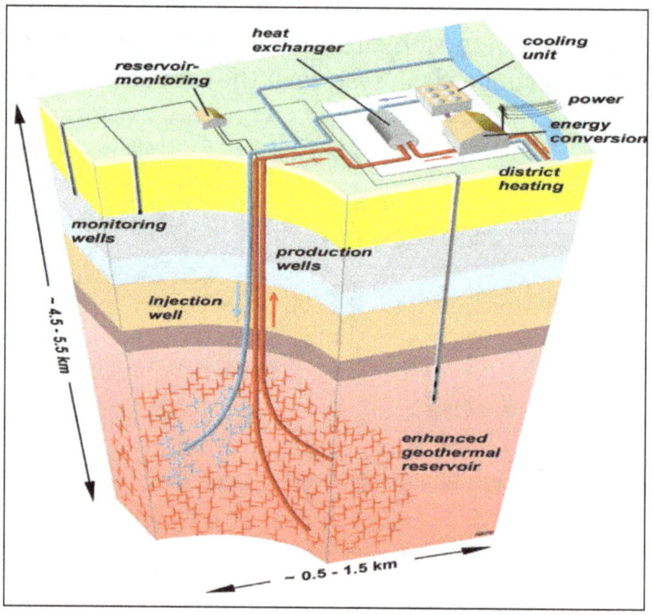

Schematic of Enhanced Geothermal System.

An Enhanced Geothermal System (EGS) also known as engineered geothermal systems is a variety of engineering techniques used to artificially create hydrothermal resources (underground steam and hot water) that can be used to generate electricity. In an EGS, fluid is injected into the subsurface under carefully controlled conditions, which cause pre-existing fractures to re-open, creating permeability. Enhanced geothermal systems utilize advanced, often experimental, drilling and fluid injection techniques to augment and expand the availability of geothermal resources, which can be used to generate electricity from the heat in the earth's crust. Enhanced geothermal systems, when recharged, can provide near continuous output, making the technology a renewable, zero-carbon option for supplying base load electricity generation.

Ground Source Heat Pump (GSHP's)

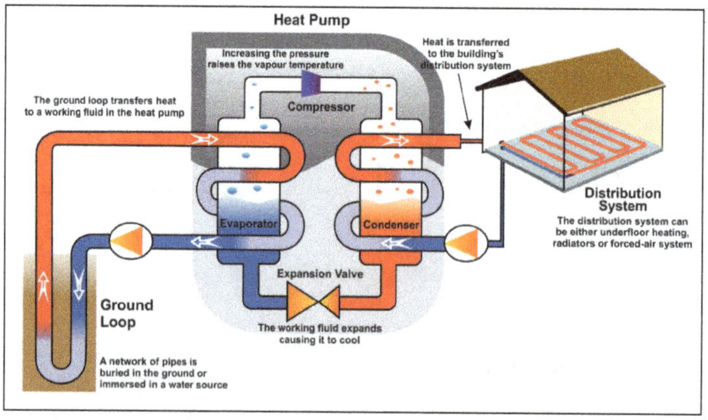

Schematic working of ground source heat pumps.

Ground Source Heat Pumps (GSHP's) use the earth's relatively constant temperature between the 16-29 °C all year round at a depth of 20 feet. The ground source heat pumps (GSHP) are also known as Ground Heat Pumps-GHP, Ground Coupled Heat Pump-GCHP, Groundwater Heat Pump-GWHP and Geo-exchange Pumps. GSHP provides heating, cooling, and hot water for homes and commercial buildings. GSHPs use the earth as a heat source (in the winter) or a heat sink (in the summer). The basic principle on which the GHP works is "refrigeration cycle". The refrigerant carries the heat from one "space" to another.

Geothermal Gradient

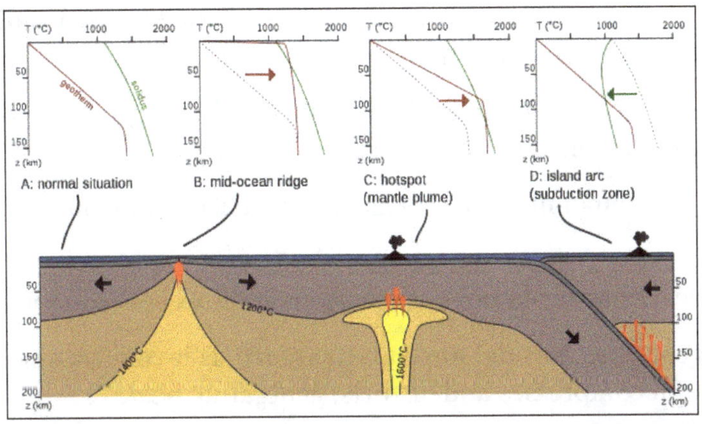

Geothermal gradient is the rate of increasing temperature with respect to increasing depth in the Earth's interior. Away from tectonic plate boundaries, it is about 25 °C per km of depth (1 °F per 70 feet of depth) in most of the world. Strictly speaking, geo-thermal necessarily refers to the Earth but the concept may be applied to other planets.

The Earth's internal heat comes from a combination of residual heat from planetary accretion; heat produced through radioactive decay, and possibly heat from other sources. The major heat-producing isotopes in the Earth are potassium-40, uranium-238, uranium-235, and thorium-232. At the center of the planet, the temperature may be up to 7,000 K and the pressure could reach 360 GPa (3.6 million atm).

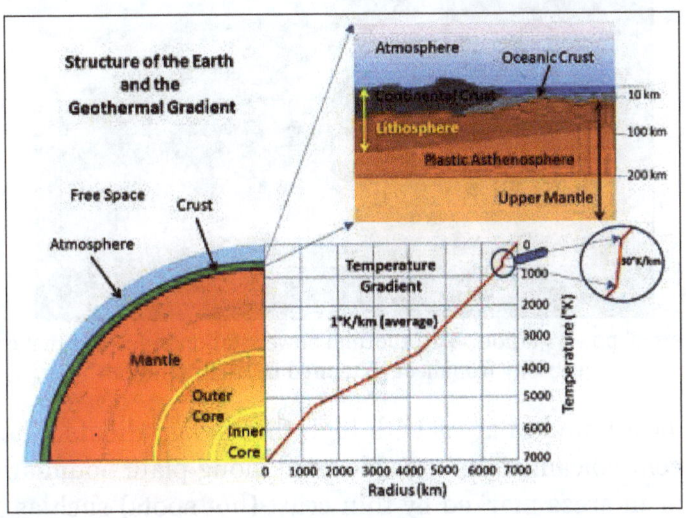

Because much of the heat is provided by radioactive decay, scientists believe that early in Earth history, before isotopes with short half-lives had been depleted, Earth's heat production would have been much higher. Heat production was twice that of present-day at approximately 3 billion years ago, resulting in larger temperature gradients within the Earth, larger rates of mantle convection and plate tectonics, allowing the production of igneous rocks such as komatiites that are not formed anymore today.

Heat Sources

- Much of the heat is created by decay of naturally radioactive elements. An estimated 45 to 90 percent of the heat escaping from the Earth originates from radioactive decay of elements mainly located in the mantle.
- Heat of impact and compression released during the original formation of the Earth by accretion of in-falling meteorites.
- Heat released as abundant heavy metals (iron, nickel, copper) descended to the Earth's core.
- Latent heat released as the liquid outer core crystallizes at the inner core boundary.
- Heat may be generated by tidal force on the Earth as it rotates; since rock cannot flow as readily as water it compresses and distorts, generating heat.
- There is no reputable science to suggest that any significant heat may be created by electromagnetic effects of the magnetic fields involved in Earth's magnetic field, as suggested by some contemporary folk theories.

Extraction, Exhaustion and Environmental Effects

Geothermal power station: A geothermal power station in Iceland that creates electricity from heat generated in Earth's interior.

Geothermal energy is best found in areas with high thermal gradients. Those gradients occur in regions affected by recent volcanism, in areas located along plate boundaries (such as along the Pacific Ring of Fire), or in areas marked by thin crust (hot spots) such as Yellowstone National

Park and the Hawaiian Islands. Geothermal reservoirs associated with those regions must have a heat source, adequate water recharge, a reservoir with adequate permeability or faults that allow fluids to rise close to the surface, and an impermeable caprock to prevent the escape of the heat. In addition, such reservoirs must be economically accessible (that is, within the range of drills).

The heated fluid from a geothermal resource is tapped by drilling wells, sometimes as deep as 9,100 metres (about 30,000 feet), and is extracted by pumping or by natural artesian flow (where the weight of the water forces it to the surface). Water and steam are then piped to the power plant to generate electricity or through insulated pipelines—which may be buried or placed aboveground—for use in heating and cooling applications. In general, electric power plant pipelines are limited to roughly 1.6 km (1 mile) in length to minimize heat loss in the steam. However, direct-use pipelines spanning several tens of kilometres have been installed with a temperature loss of less than 2–5 °C (3.6 – 9 °F), depending on the flow rate. The most economically efficient facilities are located close to the geothermal resource to minimize the expense of constructing long pipelines. In the case of electric power generation, costs can be kept down by locating the facility near electrical transmission lines to transmit the electricity to market.

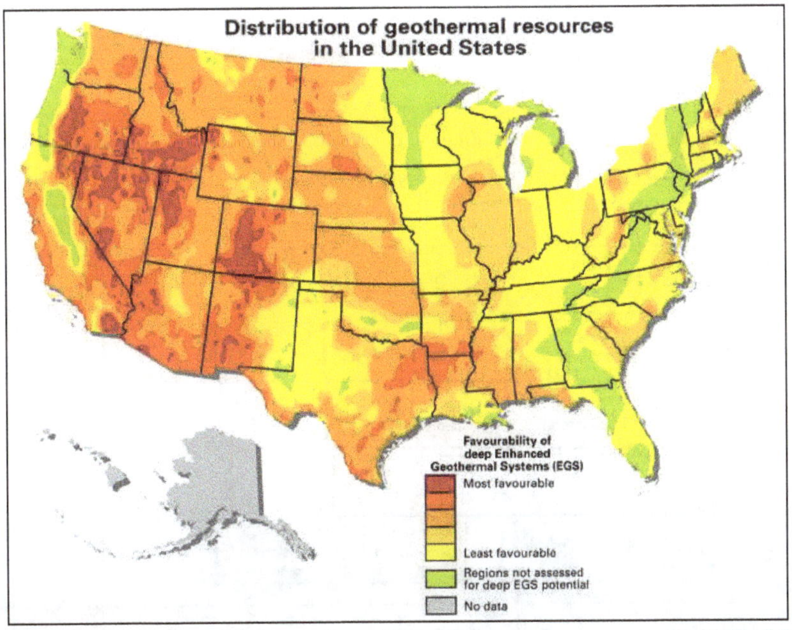

Geothermal energy potential: Map of geothermal energy resources in the United States.

Exhaustion

Geothermal resources can be exhausted if the rate of heat extraction exceeds the rate of natural heat recharge. Normally, geothermal resources can be used for 20 to 30 years; however, the energy output may decrease with time, making continued development uneconomical. On the other hand, geothermal electric power has been produced continually from the Larderello geothermal field since the early 1900s and at the Geysers since 1960. Although there has been a decline in both of those fields, this problem has been partially overcome by drilling new wells and by recharging the water supply. At the Geysers, electrical capacity declined from 1,800 MW to approximately 1,000 MW, but about 200 MW of capacity was returned by placing the field under one operator and constructing pipelines to deliver wastewater for recharging the reservoir. Projects such as the

Reykjavík district heating system have been operating since the 1930s with little change in the output, and the Oregon Institute of Technology geothermal heating system has been operating since the 1950s with no change in production. Thus, with proper management, geothermal resources can be sustainable for many years, and they can even recover if use is suspended for a period of time.

Environmental Effects and Economic Costs

The environmental effects of geothermal development and power generation include the changes in land use associated with exploration and plant construction, noise and sight pollution, the discharge of water and gases, the production of foul odours, and soil subsidence. Most of those effects, however, can be mitigated with current technology so that geothermal uses have no more than a minimal impact on the environment. For example, Klamath Falls, Oregon, has approximately 600 geothermal wells for residential space heating. The city has also invested in a district heating system and a downtown snow-melting system, and it provides heating to local businesses. However, none of the systems used to supply and deliver geothermal energy are visible in town.

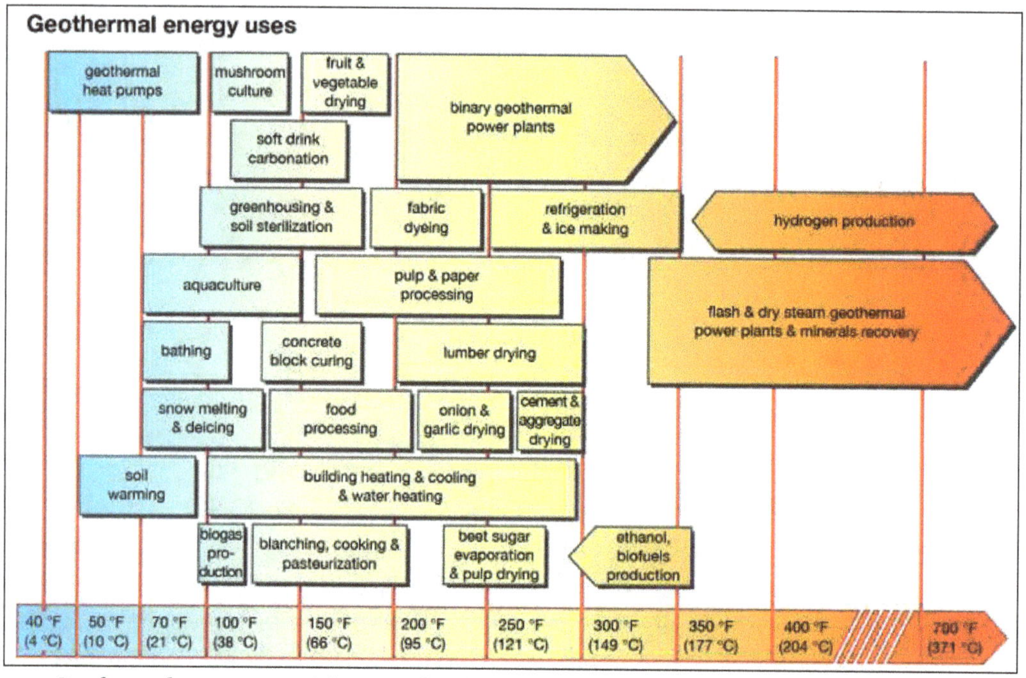

Geothermal energy uses: Diagram of various geothermal energy uses displayed according to the water temperature of the geothermal resource.

In addition, GHPs have a very minimal effect on the environment, because they make use of shallow geothermal resources within 100 metres (about 330 feet) of the surface. GHPs cause only small temperature changes to the groundwater or rocks and soil in the ground. In closed-loop systems the ground temperature around the vertical boreholes is slightly increased or decreased; the direction of the temperature change is governed by whether the system is dominated by heating (which would be the case in colder regions) or cooling (which would be the case in warmer regions). With balanced heating and cooling loads, the ground temperatures will remain stable. Likewise, open-loop systems using groundwater or lake water would have very little effect on temperature, especially in regions characterized by high groundwater flows.

Comparing the benefits of geothermal energy with other renewable energy sources, the main advantage of geothermal energy is that its base load is available 24 hours per day, 7 days per week, whereas solar and wind are available only about one-third of the time. In addition, the cost of geothermal energy varies between 5 and 10 cents per kilowatt-hour, which can be competitive with other energy sources, such as coal. The main disadvantage of geothermal energy development is the high initial investment cost in constructing the facilities and infrastructure and the high risk of proving the resources. (Geothermal resources in low-permeability rocks are often found, and exploration activities often drill "dry" holes—that is, holes that produce steam in amounts too low to be exploited economically). However, once the resource is proven, the annual cost of fuel (that is, hot water and steam) is low and tends not to escalate in price.

References

- Manzella-Geophysical-Geochemical-Meth, elearning: geothermalcommunities.eu, Retrieved 12, June 2020
- Geothermal-energy, science: britannica.com, Retrieved 13, April 2020
- Geothermal-gradient: geologyin.com, Retrieved 11, March 2020
- Environmental-effects-and-economic-costs, geothermal-energy, science: britannica.com, Retrieved 16, May 2020

Permissions

All chapters in this book are published with permission under the Creative Commons Attribution Share Alike License or equivalent. Every chapter published in this book has been scrutinized by our experts. Their significance has been extensively debated. The topics covered herein carry significant information for a comprehensive understanding. They may even be implemented as practical applications or may be referred to as a beginning point for further studies.

We would like to thank the editorial team for lending their expertise to make the book truly unique. They have played a crucial role in the development of this book. Without their invaluable contributions this book wouldn't have been possible. They have made vital efforts to compile up to date information on the varied aspects of this subject to make this book a valuable addition to the collection of many professionals and students.

This book was conceptualized with the vision of imparting up-to-date and integrated information in this field. To ensure the same, a matchless editorial board was set up. Every individual on the board went through rigorous rounds of assessment to prove their worth. After which they invested a large part of their time researching and compiling the most relevant data for our readers.

The editorial board has been involved in producing this book since its inception. They have spent rigorous hours researching and exploring the diverse topics which have resulted in the successful publishing of this book. They have passed on their knowledge of decades through this book. To expedite this challenging task, the publisher supported the team at every step. A small team of assistant editors was also appointed to further simplify the editing procedure and attain best results for the readers.

Apart from the editorial board, the designing team has also invested a significant amount of their time in understanding the subject and creating the most relevant covers. They scrutinized every image to scout for the most suitable representation of the subject and create an appropriate cover for the book.

The publishing team has been an ardent support to the editorial, designing and production team. Their endless efforts to recruit the best for this project, has resulted in the accomplishment of this book. They are a veteran in the field of academics and their pool of knowledge is as vast as their experience in printing. Their expertise and guidance has proved useful at every step. Their uncompromising quality standards have made this book an exceptional effort. Their encouragement from time to time has been an inspiration for everyone.

The publisher and the editorial board hope that this book will prove to be a valuable piece of knowledge for students, practitioners and scholars across the globe.

Index